CFD Techniques and Thermo-Mechanics Applications

Zied Driss · Brahim Necib
Hao-Chun Zhang
Editors

CFD Techniques and Thermo-Mechanics Applications

Springer

Editors
Zied Driss
Department of Mechanical Engineering
National School of Engineers of Sfax
Sfax
Tunisia

Hao-Chun Zhang
School of Energy Science and Engineering
Harbin Institute of Technology
Harbin
China

Brahim Necib
Faculty of Sciences and Technology
University of Constantine 1
Constantine
Algeria

ISBN 978-3-319-70944-4 ISBN 978-3-319-70945-1 (eBook)
https://doi.org/10.1007/978-3-319-70945-1

Library of Congress Control Number: 2017963845

Printed on acid-free paper

This Springer imprint is published by the registered company Springer International Publishing AG part of Springer Nature
The registered company address is: Gewerbestrasse 11, 6330 Cham, Switzerland

Preface

This book focuses on Computational Fluid Dynamics (CFD) technics and the recent developments and research works in thermomechanics applications. It is also devoted to the publication of basic and applied studies broadly relating to this area. The chapters may present the development of numerical methods, computational techniques, and case studies in the thermomechanics applications. Also, they offer the fundamental knowledge for using CFD in real thermomechanics applications and complex flow problems through the new technical approaches. It describes the steps in the CFD process and provides benefits and issues when using the CFD analysis in understanding of complicated flow phenomena and its use in the design process. The best practices for reducing errors and uncertainties in CFD analysis may be adapted. The presented case studies and developments approaches aim to provide the readers, such as engineers and Ph.D. students, the fundamentals of CFD prior to embarking on any real simulation project. Additionally, engineers supporting or being supported by CFD analysts can benefit from this book.

Chapter "Air Flow CFD Modeling in an Industrial Convection Oven" discusses an experimentally validated 3-D CFD analysis of the flow and thermal processes in a laboratory drying oven with a forced air circulation. The thermal field within an oven has significant impact on quality of cooked food, and reliable predictions are important for a robust design and performance evaluation of an oven. A numerical simulation by using a computational fluid dynamics code is carried out to predict the three-dimensional isothermal airflow in an industrial electrical forced convection oven. The CFD model is based on the fundamental equations for the conservation of mass, momentum, and the k-ε turbulence model. The CFD model performance was assessed by means of point measurements of the velocity with a directionally hot-film velocity sensor. The simulated results were consistent with the actual velocity measurements from the industrial oven.

Chapter "CFD Application for the Study of Innovative Working Fluids in Solar Central Receivers" focuses on the evaluation of a new HTF (supercritical CO_2) used in a solar central receiver in comparison with a commercial one (molten salt) using a CFD model. In this chapter, the results related to the operating conditions for the innovative HTF and to the adaptation of the solar-receiver design have been discussed and analyzed. In fact, Concentrating Solar Thermal (CST) technologies are focused on the production of both electricity and heat by the concentration of sunlight direct-beam part. Thus, Solar Thermal Electricity (STE) plants collect and concentrate the solar energy which is converted into heat by using a Heat Transfer Fluid (HTF) in the solar receiver, and, in the second step, the heat is transformed into electricity by a power block. The selection of an appropriate HTF is important for increasing both the efficiency of the solar receiver and the overall efficacy of the STE plant.

In Chapter "Computational Fluid Dynamics for Thermal Evaluation of Earth-to-Air Heat Exchanger for Different Climates of Mexico", a two-dimensional model based on computational fluid dynamics is developed to analyze the thermal performance of an Earth-to-Air Heat Exchanger (EAHE) in three cities of Mexico. The climatic data correspond to a temperate climate (México City), a humid–hot climate (Mérida, Yucatán), and an extreme weather (Juárez City, Chihuahua). The optimal depth of burial of the EAHE for the three cities has been found. The temperature, velocity, and cooling variation and the heating potential for each case of study were presented. The results show that the cooling and cooling potential change with the depth of burial of the tube.

In Chapter "CFD Modeling of a Parabolic Trough Receiver of Different Cross Section Shapes", the Parabolic Trough Collector (PTC) performance was examined. In order to reach this aim, the adopted method comprises two major steps. In the first step, the concentrated solar heat flux densities in a focal zone are calculated by SOLTRACE software. In the second step, some Computational Fluid Dynamics (CFD) simulations are carried out to analyze and optimize the thermal performance of the tube receiver. The calculated heat flux densities by SOLTRACE software are used as heat flux wall boundary conditions for the receiver tube. The effect of the absorber tube cross-sectional shape on the performance of the PTC system is analyzed. Triangular, rectangular, and circular shapes are tested, and the results are compared.

In Chapter "An OpenFOAM Solver for Forced Convection Heat Transfer Adopting Diagonally Implicit Runge Kutta Schemes", a CFD solver was developed for incompressible fluid flow and forced convection heat transfer based on high-order diagonally implicit Runge–Kutta (RK) schemes for time integration. In particular, an iterated PISO-like procedure based on Rhie–Chow correction was used for handling pressure–velocity coupling within each RK stage. It is worth emphasizing that for space discretization the numerical technology available within the well-known OpenFOAM library was used. The aim of this chapter is to explore the reliability and effectiveness of OpenFOAM library for convective heat transfer problems using high-fidelity numerics.

In Chapter "Multigrid and Preconditioning Techniques in CFD Applications", multigrid and preconditioning techniques allowing to speed up CFD calculations on unstructured meshes are discussed. Flow solution is provided using cell-centered finite volume formulation of unsteady three-dimensional compressible Navier–Stokes equations on unstructured meshes. The CFD code uses an edge-based data structure to give the flexibility to run on meshes composed of a variety of cell types.

The fluxes are calculated on the basis of flow variables at nodes at either end of an edge or an area associated with that edge (edge weight). The edge weights are precomputed and take into account the geometry of the cell. The capabilities of the approaches developed are demonstrated by solving benchmark problems on structured and unstructured meshes.

Chapter "Numerical Simulation and Experimental Validation of the Role of Delta Wing Privileged Apex" is devoted to the numerical and experimental studies of a thin delta wings aerodynamics with "Privileged Angles". This study focuses on observations and visualizations in the wind tunnel. It suggested that the delta wings with "privileged" apex can influence the delta wing aerodynamic characteristics and consequently could have repercussions on the aircraft performances. In addition, it revealed that the apex vortex which develops on the suction face of this type of wings occupies positions corresponding to values of quantified angles, called "Privileged Angles". The delta wing vortex lift is mainly due to the depression generated on its extrados part (suction face) by a three-dimensional (3D) flow resulting from the complex swirling structure, which occurs at the leading edge of the wing. Relatively, the topology of this type of flow is well known, but the character of the mechanism remains to specify.

Chapter "Numerical Simulation of the Overlap Effect on the Turbulent Flow Around a Savonius Wind Rotor" aims to investigate the effect of the overlap on the aerodynamic characteristics of the flow around a Savonius wind rotor. Thus, the writers have developed a numerical simulation using a commercial CFD code. The considered numerical model is based on the resolution of the Navier–Stokes equations in conjunction with the k-ε turbulence model. These equations are solved by a finite volume discretization method. The comparison of the numerical results with anterior results shows a good agreement.

Chapter "Study of the Collector Diameter Effect on the Characteristics of the Solar Chimney Power Plant" aims to optimize the geometry of the collector in a Solar Chimney Power Plant (SCPP). Particularly, the effects of the collector diameter on the SCPP output are investigated. A two-dimensional steady model with the standard k-ε turbulence model has been developed using the commercial Computational Fluid Dynamics (CFD) code "ANSYS Fluent 17.0". A numerical simulation was performed to study the local characteristics of the air flow inside the SCPP. The local flow characteristics were presented and discussed for different collector diameters. The comparison shows that the collector diameter is an important parameter for the optimization of the solar setup.

Sfax, Tunisia — Zied Driss
Constantine, Algeria — Brahim Necib
Harbin, China — Hao-Chun Zhang

Acknowledgements

First and foremost, I would like to thank Dr. Nabil Khélifi, Springer Editor who invited me to edit this new book after awarding the conference on CFD techniques and Thermo-Mechanics applications, which was held at the National School of Engineers of Sfax (University of Sfax, Tunisia) in April 2016. All the ideas have developed further with my co-editors and many reviewers, especially in the second edition of the International Conferences on Mechanics and Energy (ICME'2016) which was held in Hammamet (Tunisia) in December 2016 and the third edition ICME'2017, held in Sousse (Tunisia) in December 2017.

I would like to thank all the authors who submitted chapters at our requests. Especially, I wish to express my gratitude to all the reviewers who participated to this book, provided support, talked things over, read, wrote, offered comments, and allowed us to quote their remarks.

Many colleagues have generously provided comments and material from their past and current research. Particularly, I thank my co-editors Prof. Brahim Necib from the University of Mentouri Constantine (Algeria) and Prof. Hao-Chun Zhang from the Harbin Institute of Technology (China). Without them, this book would never find its way to so many researchers, engineers, and Ph.D. students.

I would like to express my gratitude to all those who provided support and assisted in the editing and proofreading. Particularly, I thank Prof. Abdelmajid Dammak for the Linguistic improvements of all chapters in the book. In addition, I would like to thank Reyhaneh Majidi, Shahid Mohammed, Kavitha Palanisamy, and Suganya Manoharan from Springer for helping me in the process of selection, editing, and design.

Last and not least, I beg forgiveness of all those who have been with me over the course of the years and whose names I have failed to mention.

Sfax, Tunisia
January 2018

Prof. Dr. Zied Driss

Contents

Air Flow CFD Modeling in an Industrial Convection Oven

Julio Cesar Zanchet Piaia, Carlos Alberto Claumann, Marintho Bastos Quadri and Ariovaldo Bolzan

1 Introduction

Owing to the changing consumer habits and, correspondingly, the increased consumption of ready meals, food reheating operations are becoming increasingly important. The aim of such processes is usually to achieve a maximum temperature uniformity inside the food as fast as possible.

Drying ovens and thermostatic cabinets are broadly used in installations and also in laboratories in food, pharmaceutical, chemical and paper industries, etc., for storing products at constant temperature conditions (Smolka et al. 2010). The operation principle of these ovens is based on the circulation of hot air inside the oven and convective heat transfer from the air to the food surface (Stigter et al. 2001).

The thermo-fluid dynamics within an oven has a significant impact on the quality of cooked food and reliable predictions are important for a robust design and performance evaluation of an oven (Mistry et al. 2006). In order to achieve a better design, the heat transfer mechanisms inside the oven must be well understood. Due to complexity involved in the flow dynamics, a modeling approach may be appropriate in order to target and reduce experiments. In fact, a validated model

J. C. Z. Piaia (✉) · C. A. Claumann · M. B. Quadri · A. Bolzan
Chemical and Food Engineering Department, Santa Catarina Federal University, Mail Box: 476, Florianópolis, SC 88040-900, Brazil
e-mail: julio.piaia@posgrad.ufsc.br

C. A. Claumann
e-mail: carlos.claumann@prosgrad.ufcs.br

M. B. Quadri
e-mail: m-quad@enq.ufsc.br

A. Bolzan
e-mail: abolzan@enq.ufsc.br

Z. Driss et al. (eds.), *CFD Techniques and Thermo-Mechanics Applications*,
https://doi.org/10.1007/978-3-319-70945-1_1

allows parametric phenomenon studies without additional operating costs and within a short period of time (Mirade et al. 2002).

The computational approaches aiming at improving the process and/or device design very often employ a computational fluid dynamics (CFD). This technique was also successfully applied for the device improvements in the food industry. In the food industry, one of the most important and successfully applied techniques is CFD which is capable of visualizing a three-dimensional flow and temperature field in the considered domains. Particularly, the CFD was used to design and optimize the humidification process of cold stores (Delele et al. 2009), a continuous bread baking (Wong et al. 2007a, b), the ripening of fermented food products (Mirade 2008), and drying process of figures (Stamatios et al. 2004).

Smolka et al. (2013) formulated and experimentally validated a numerical CFD model of a heating oven with natural hot-air circulation. The problem was solved using a three-dimensional accurate geometry and the commercial CFD package. Based on the developed model, the most important parameters and the device parts that influence the temperature uniformity were determined. Additionally, two potential modifications to optimize the heating device have been suggested.

Khatir et al. (2012) investigated the air flow and temperature distribution in a small-scale bread-baking oven using a CFD model. The model has been validated against experimental data and has shown to be capable of providing valuable insight into key baking issues such as temperature uniformity that are difficult and time-consuming to measure experimentally. The CFD results demonstrate that careful selection of the flow model, together with implementation of realistic boundary conditions, give accurate temperature predictions throughout the oven.

Therdthai et al. (2004) developed a three-dimensional computational fluid dynamics (CFD) model to simulate the temperature profiles and airflow patterns during a continuous industrial baking process. The model was further used to investigate the oven operating conditions which could produce the optimum baking condition. According to the simulation results, the heat supply could be reduced whereas the airflow volume should be increased. Thanks to this modification, the bread weight loss could be reduced by 1.4% while keeping the same characteristics.

Mirade et al. (2004) CFD approach was applied to predict the air temperature and velocity profiles in an industrial biscuit baking tunnel oven. They used two three-dimensional CFD models: one not covering the conveying band of biscuits and the other including it to describe the complex air circulation resulting from the mechanisms of air input and exit at the ends of the oven and of air extraction through the different extraction points located along the oven length. Comparing the numerical results to the experimental measurements, the authors report satisfactory results for the phenomena under study.

The objective of this research was to develop and validate a CFD model of air forced convection oven. This study discusses the major aspects of the airflow model and its validation through a hot-film velocity sensor located in 48 different points of the oven cavity.

2 Materials and Methods

2.1 Forced Convection Oven

The oven used in this study is an industrial electrical forced convection oven device for the heating of foods. The oven contains eight grid shelves, with support for removable trays. Each grid shelf is 59 cm wide and 65 cm deep. The total dimensions of the useful oven cavity are 70 cm wide, 84 cm deep, and 90 cm high. The external dimensions of the oven, taking into account the wall insulation, are 98 cm wide, 100 cm deep, and 130 cm in height, as shown in Fig. 1.

There is a fan fixed at the backside wall of the oven with a diameter of 35 cm. A plug fan sucks the hot air out of the processing area axially and blows toward the walls in circular movement. The air passes through tubes of 2 cm in diameter and 76 cm long, placed vertically inside the walls of the cavity. The heating is achieved by the means of coils inside the tube and is then distributed throughout the oven. The radial fan is connected to a frequency inverter for the speed control system and rotates counterclockwise direction with a maximum speed of 3465 rpm. In this study, we used a rotation speed of 1730 rpm with the vacuum oven, in order to simplify the model validation.

2.2 Governing Equations

The equations, which describe the conservation of mass and momentum of a moving incompressible Newtonian fluid such as air, are written in index notation for Cartesian coordinates. For a three-dimensional turbulence flow, isothermal process at stationary state can be mathematically described by the following differential equations (Bird et al. 2002):

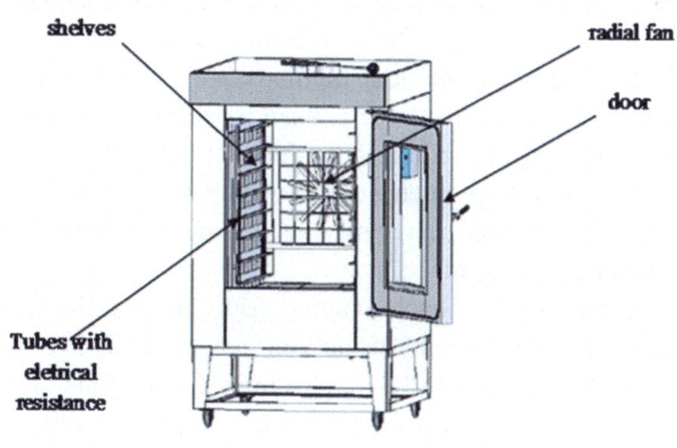

Fig. 1 Illustration of the electrical forced convection oven

Mass conservation (Continuity equation):

$$\frac{\partial u}{\partial x}+\frac{\partial v}{\partial y}+\frac{\partial w}{\partial z}=0 \tag{1}$$

Momentum conservation (Navier–Stokes equation):

$$\rho\left(u\frac{\partial u}{\partial x}+v\frac{\partial u}{\partial y}+w\frac{\partial u}{\partial z}\right)=-\frac{\partial p}{\partial x}+\mu\left(\frac{\partial^2 u}{\partial x^2}+\frac{\partial^2 u}{\partial y^2}+\frac{\partial^2 u}{\partial z^2}\right)+f_X \tag{2}$$

$$\rho\left(u\frac{\partial v}{\partial x}+v\frac{\partial v}{\partial y}+w\frac{\partial v}{\partial z}\right)=-\frac{\partial p}{\partial y}+\mu\left(\frac{\partial^2 v}{\partial x^2}+\frac{\partial^2 v}{\partial y^2}+\frac{\partial^2 v}{\partial z^2}\right)+f_Y \tag{3}$$

$$\rho\left(u\frac{\partial w}{\partial x}+v\frac{\partial w}{\partial y}+w\frac{\partial w}{\partial z}\right)=-\frac{\partial p}{\partial z}+\mu\left(\frac{\partial^2 w}{\partial x^2}+\frac{\partial^2 w}{\partial y^2}+\frac{\partial^2 w}{\partial z^2}\right)+f_Z, \tag{4}$$

where p is the static pressure (N/m^2), v is the velocity component (m/s), ρ is density (kg/m^3), f is the inner volumetric mass generation rate (kg/sm^3), and μ is the dynamic viscosity (Pa s).

2.3 CFD Simulations

The commercial CFD software, Ansys CFX 11.0 was used for discretization and solution of equations. This commercial package consists of three modules: the preprocessor, the processor, and the postprocessor. The preprocessor consists of two parts: the first part is used to build a multiblock geometry, composed of solid hexahedron, where it supports the construction of tetrahedral mesh number on solids surfaces (computational geometry representation in the plan). The second part is used to generate a file where the mathematical models are chosen generating a numerical model. The processor is a module of numerical solutions, where the finite volumes numerical methods are used to solve generated model in preprocessor. The postprocessor is used for scientific visualization of the results obtained in the processor module in the form of color maps, vector graphics, and animations.

The 3D full geometry of the oven was accurately reproduced. The internal walls, door, pipe, radial fan, and shelves are represented by hexahedral meshes in the computational grid.

In the center of each existing grid volumes, the software calculates the mass and momentum balance by finite volumes method. The discretization scheme used was UPWIND (Maliska 2004). Discretized equations were solved iteratively using the SIMPLEC algorithm for pressure–velocity coupling and the solution was considered converged when the total residuals for the continuity equation, scaled by its largest absolute value over the first five iterations, dropped to below 10^{-4}.

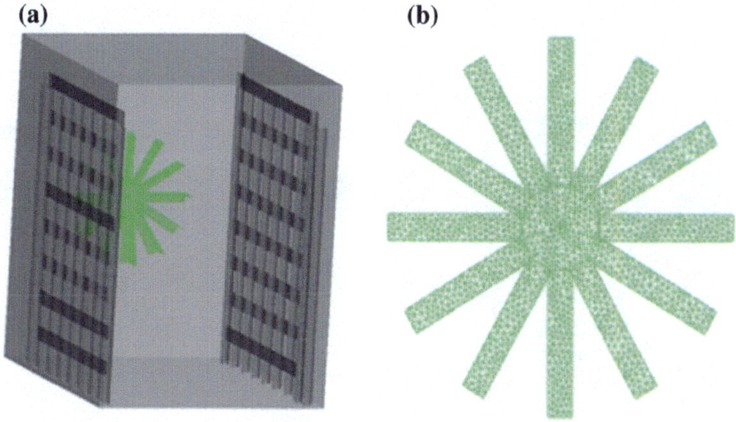

Fig. 2 Generated geometry: **a** internal camber, **b** grid view of radial fan (tetrahedric mesh)

Model assumptions and boundary conditions are turbulence flow, steady state, three-dimensional flow, and isothermal. The model accepts the following hypotheses and boundary conditions: turbulent flow, steady state, three-dimensional and isothermal flow. Zero flow velocities (no-slip condition) were set in every wall, and air properties were taken as constants and calculated for the reference temperature of 25 °C, density ρ_a equals 1.21 (kg/m^3), and dynamic viscosity equals 1.81×10^{-5} kg/ms. Three different grid sizes were studied, a poorly refined grid made of 919,198 volume elements, an intermediate one with 1,438,127 volumes, and a refined grid with 2,682,371 volumes.

The generated geometry of the oven internal chamber and the grid view of the radial fan can be seen in Fig. 2.

A stationary flow field was achieved after 900 time steps using ten outer iterations for each time step. The model was run on a Core 2 Quad Q6600 2.4 GHz processor and 8 GB RAM. The total computation time was approximately 5 h with the finest grid.

2.4 Validation Method

Three hot-film anemometer sensors (EE65/EE Elektronik) with an operating range of (0–20 m/s) were used to monitor the average speeds. In usual practice, the sensor cover is used to measure the two-dimensional flow perpendicular to the wire (accuracy ±2% of reading).

One of the most frequently used techniques to perform velocity measurements in fluids is hot-film anemometry (Sanyal et al. 2006). This technique is based on the relationship between the heat flux from an electrically heated sensing element, which is placed in a flow, and the local fluid velocity. The heat exchange rate

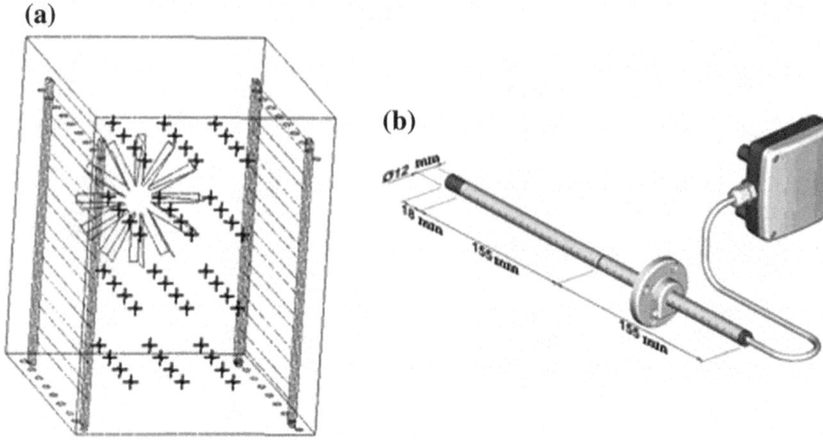

Fig. 3 Experimental method: **a** location of the velocity measurements, **b** hot-film anemometer used

between the filament and the environment is a speed-flow function, thus speed values can be obtained.

Figure 3 illustrates the experimental methodology for monitoring the average speeds in the oven cavity.

The velocity transducer is connected to a portable PC with a data acquisition FIELDLOGGER (Novus, Brazil). The software FIELDCHART (Novus, Brazil) was used to interface with the data acquisition. 48 points in the oven cavity were scanned (12 points on four of the eight shelves), as illustrated in Fig. 3. In each point, the orthogonal direction throughout the fan was measured separately, over a sufficiently large period (120 s).

The CFD modeling error is calculated as the mean rate (in absolute value) between the differences of experimental velocities and the predicted ones by the model

$$P = \frac{100}{n} \cdot \sum \frac{\left| v_{\mathrm{CFD}} - v_{\exp} \right|}{v_{\exp}}, \tag{5}$$

where $v_{\exp}$ is the experimental velocity in the total time of 120 s, v_{CFD} is the velocity prediction model, and n is the number of the experimental data.

3 Results and Discussion

The velocity profile between the shelves was investigated using the CFD techniques. The CFD velocity profile calculations were compared with the measured data at each point of the oven cavity with a period of 120 s.

The details of the velocity validation are now discussed. The main aspects of the calculated flow are illustrated in Fig. 4, which shows the velocity vectors in a horizontal cross section of the oven model. Due to the complex geometry and the swirl of the fan, the air does not travel in a straightforward way through the oven cavity toward the fan. The radial fan directs the airflow toward the walls located nearby the tubes with an electrical resistance. The air returns to the fan through the center of the oven, and it is at this cavity where the food to be processed is located.

A qualitative and quantitative comparison of the measurements and the calculations for the velocity components in the orthogonal direction *y* are given in Figs. 5, 6, 7, and 8. The measured and predicted velocity magnitudes are compared on different horizontal lines in the oven cavity, the coordinate *y* from the left side, center, and the right side of the oven cavity. Results for the three grids sizes are plotted.

The highest velocities observed in the oven are found nearby the fan. Figure 6 shows that the modeling error is higher when compared to the velocities near the door.

Table 1 displays the average errors (*P*) and the computation times obtained by CFD computational model with different mesh sizes in comparison with measurements.

As a result, some errors were produced from the simplification of the geometry of forced convection oven air. As shown in Table 1, when CFD model measurements average velocities are compared, the smallest error obtained is 18.14%, using the largest number of volume elements, 2,682,371. The error was reduced by increasing the grid refinement at the cost of increasing the simulation time. Due to the limitation of the available computer resources, the grid could not be further refined.

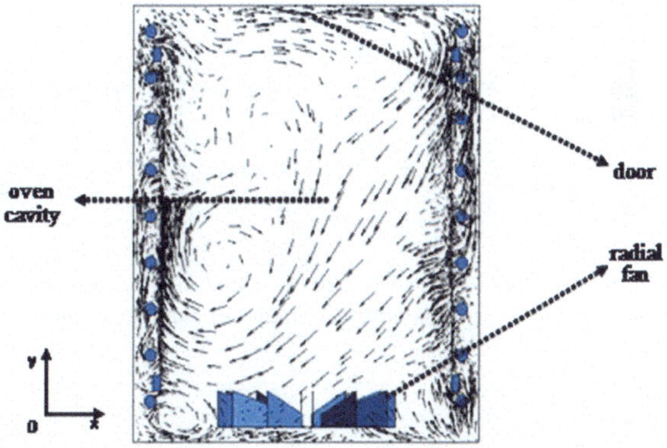

Fig. 4 Calculated velocity vectors in a horizontal cross section of the oven (shelf 8)

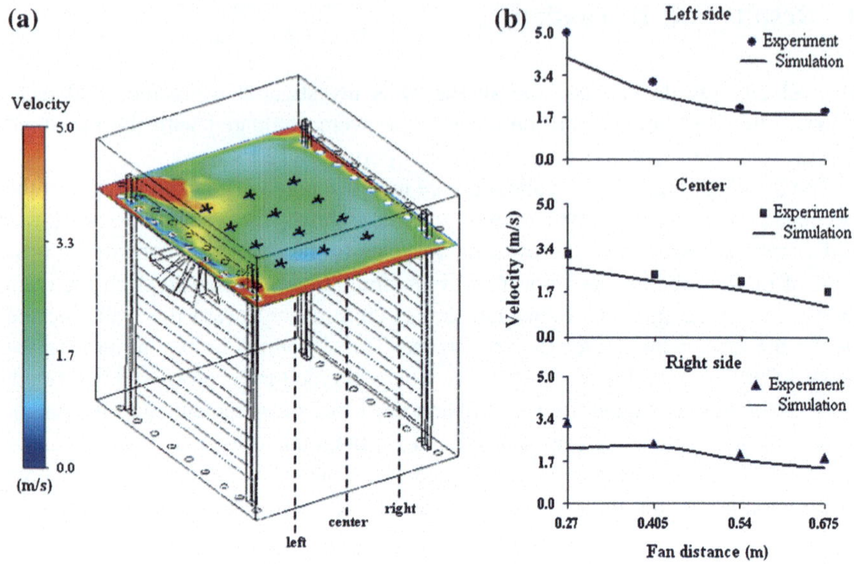

Fig. 5 Air velocity profile, shelf 2: **a** horizontal cross section, **b** comparison of total velocity magnitude to experimentally measured and simulated values

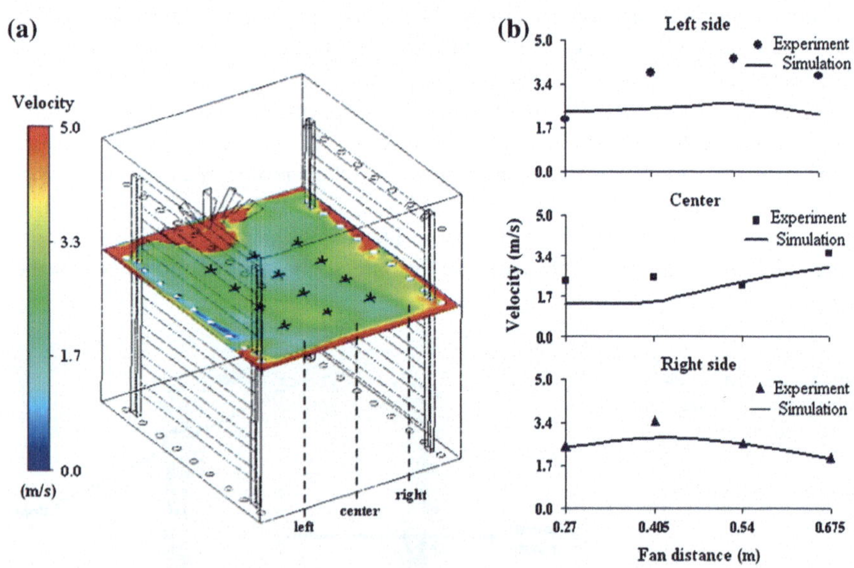

Fig. 6 Air velocity profile, shelf 4: **a** horizontal cross section, **b** comparison of total velocity magnitude to experimentally measured and simulated values

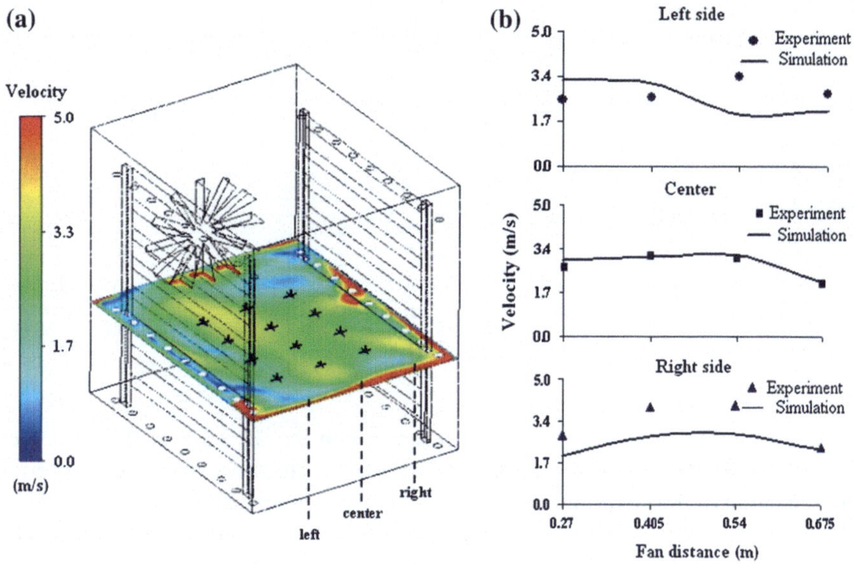

Fig. 7 Air velocity profile, shelf 6: **a** horizontal cross section, **b** comparison of total velocity magnitude to experimentally measured and simulated values

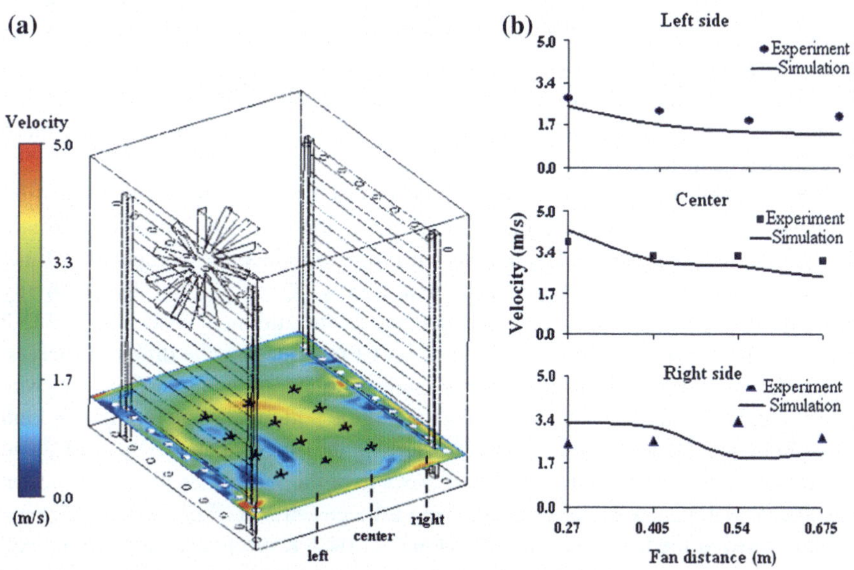

Fig. 8 Air velocity profile, shelf 8: **a** horizontal cross section, **b** comparison of total velocity magnitude to experimentally measured and simulated values

Table 1 CFD calculation error and computational time for different grids

Number of volumes	P (%)	CPU (min)
919,198	30.47	173
1,438,127	22.32	187
2,682,371	18.14	324

Table 2 Distribution of the CFD calculation error over the different shelves

Shelf	v_{exp} (m/s)	v_{CFD} (m/s)	P (%)
2	2.56	2.13	16.13
4	2.88	2.25	21.84
6	2.99	2.67	21.04
8	2.72	2.52	13.55

The error decreased approximately linearly with the decrease of the grid size, thus it can be stated that the error is to a large extent a result of the numerical discretization of the governing equations, which have a limited accuracy because of the first-order approximation of the convection terms. Higher order schemes sometimes produce unphysical results and increase the computational time (Verboven et al. 2000).

The magnitude of this error is in line with those reported in literature on CFD studies. Therdthai et al. (2004) reported an average relative error of 37.31% comparing the modeled velocity to the measured one during a continuous industrial baking process. Hoang et al. (2000) observed an error of 26–28.5%, using a CFD model to simulate the airflow pattern in a cold store with pallet boxes. Verboven et al. (2000) obtained an error of 22% to predict the airflow in a forced convection oven. Mirade and Daudin (1998) reported an error of 40% to predict airflow pattern in a chiller with objects.

The turbulent regime was adopted in accordance with the model suggested by Wong et al. (2007a, b) the k-ε model, this model introduces two new variables in the Navier–Stokes system of equations, in order to reduce the computation time. These two flow properties, turbulent kinetic energy (k) and the turbulence energy dissipation rate (ε) per unit mass are defined as the turbulent viscosity. Table 2 gives the distribution of the error obtained by CFD model for the different shelves.

As shown in Table 2, there is a slight increase in the error in middle of the oven (shelf 4 and 6), with the best predictions toward the bottom and the top. In this kind of equipment, the gas enters from the side of the fan wheel, reaches 90°, and accelerates due to centrifugal force as it flows over the fan blades and exits the fan housing, which can generate pressure rise in the gas stream. Shelves 4 and 6, near the fan, showed the highest errors: 21.84 and 21.04% respectively showing the bad performance of the model in predicting the correct air velocities in high turbulence areas.

According to Shih et al. (2008) and Bredell et al. (2006) in order to better predict the behavior of a fan, a mesh refinement has to be performed, increasing the number of volumes in regions close to the blades.

4 Conclusion

A three-dimensional CFD model was developed to simulate the air flow dynamics in a forced convection oven. The air velocities at steady state obtained by the computer model were compared with to the experimental measurements by means of average relative errors. Good agreement between the measurements and calculations could be observed. The average error remains 18.14% of the actual velocity, this error was caused by the limitations in turbulence modeling and numerical grid density.

The CFD application to prediction problems is not a novelty in the scientific community. Researchers of several fields have applied computational fluid dynamics with good results to problems aiming to reproduce flow patterns. In the future, the model will be extended to studying heat and mass transfer phenomena on food products subjected to cooking.

References

Bird RB, Stewart WE, Lightfoot EN (2002) Transport phenomena, 2nd edn. Wiley, New York

Bredell JR, Kröger DG, Thiart GD (2006) Numerical investigation of fan performance in a forced draft air-cooled steam condenser. Appl Therm Eng 26:846–856. https://doi.org/10.1016/j.applthermaleng.2005.09.020

Delele MA, Schenk A, Tijskens E, Ramon H, Nicolai BM, Verboven P (2009) Optimization of the humidification of cold stores by pressurized water atomizers based on a multiscale CFD model. J Food Eng 91:228–239. https://doi.org/10.1016/j.jfoodeng.2008.08.027

Hoang ML, Verboven P, De Baerdemaeker J, Nicolaï BM (2000) Analysis of the air flow in a cold store by means of computational fluid dynamics. Int J Refrig 23:127–140. https://doi.org/10.1016/S0140-7007(99)00043-2

Khatir Z, Paton J, Thompson H, Kapur N, Toropov V, Lawes M, Kirk D (2012) Computational fluid dynamics (CFD) investigation of air flow and temperature distribution in a small scale bread-baking oven. Appl Energ 89(1):89–96. https://doi.org/10.1016/j.apenergy.2011.02.002

Maliska CR (2004) Transferência de calor e mecânica dos fluidos computacional. In: LTC (ed), 2nd edn. Rio de Janeiro

Mirade PS (2008) Computational fluid dynamics (CFD) modelling applied to the ripening of fermented food products: basics and advances. Trends Food Sci Tech 19:472–481. https://doi.org/10.1016/j.tifs.2008.01.014

Mirade PS, Daudin JD (1998) Numerical simulation and validation of the air velocity field in a meat chiller. Internat J Appl Sci Comput 5(1):11–24

Mirade PS, Daudin JD, Ducept F, Trystram G, Clément J (2004) Characterization and CFD modelling of air temperature and velocity profiles in an industrial biscuit baking tunnel oven. Food Res Int 37:1031–1039. https://doi.org/10.1016/j.foodres.2004.07.001

Mirade PS, Kondjoyan A, Daudin JD (2002) Three-dimensional CFD calculations for designing large food chillers. Comput Electron Agr 34:67–88. https://doi.org/10.1016/s0168-1699(01)00180-6

Mistry H, Dey S, Bishnoi P, Castillo JL (2006) Modeling of transient natural convection heat transfer in electric ovens. Appl Therm Eng 26:2448–2456. https://doi.org/10.1016/j.applthermaleng.2006.02.007

Sanyal N, Bhattacharyya B, Munshi S (2006) An analog non-linear signal conditioning circuit for constant temperature anemometer. Measurement 39(4):308–311. https://doi.org/10.1016/j.measurement.2005.11.020

Shih YC, Hou HC, Chiang H (2008) On similitude of the cross flow fan in a split-type airconditioner. Appl Therm Eng 28:1853–1864. https://doi.org/10.1016/j.applthermaleng.2007.11.022

Smolka J, Bulinski Z, Nowak AJ (2013) The experimental validation of a CFD model for a heating oven with natural air circulation. Appl Therm Eng 54(2):387–398. https://doi.org/10.1016/j.applthermaleng.2013.02.014

Smolka J, Nowak AJ, Rybarz D (2010) Improved 3-D temperature uniformity in a laboratory drying oven based on experimentally validated CFD computations. J Food Eng 97(3):373–383. https://doi.org/10.1016/j.jfoodeng.2009.10.032

Stamatios J, Babalis SJ, Belessiotis VG (2004) Influence of the drying conditions on the drying constants and moisture diffusivity during the thin-layer drying of figs. J Food Eng 65:449–458. https://doi.org/10.1016/j.jfoodeng.2004.02.005

Stigter JD, Scheerlinck N, Nicolaï B, Van Impe JF (2001) Optimal heating strategies for a convection oven. J Food Eng 48:335–344. https://doi.org/10.1016/S0260-8774(00)00176-X

Therdthai N, Zhou W, Adamczak T (2004) Three-dimensional CFD modeling and simulation of the temperature profiles and airflow patterns during a continuous industrial baking process. J Food Eng 65:599–608. https://doi.org/10.1016/j.jfoodeng.2004.02.026

Verboven P, Scheerlinck N, De Baerdemaeker J, Nicolaï BM (2000) Computational fluid dynamics modeling and validation of the isothermal airflow in a forced convection oven. J Food Eng 43:41–53. https://doi.org/10.1016/S0260-8774(99)00131-4

Wong SY, Zhou W, Hua J (2007a) CFD modeling of an industrial continuous bread-baking process involving U-movement. J Food Eng 78:888–896. https://doi.org/10.1016/j.jfoodeng.2008.08.027

Wong SY, Zhou W, Hua J (2007b) Designing process controller for a continuous bread baking process based on CFD modeling. J Food Eng 81:523–534. https://doi.org/10.1016/j.jfoodeng.2006.12.006

CFD Application for the Study of Innovative Working Fluids in Solar Central Receivers

María Isabel Roldán Serrano, Jesús Fernández Reche and Eduardo Zarza Moya

1 Introduction

Solar thermal electricity (STE) industry has experienced important growth since 2009 and concentrating solar thermal (CST) technologies have emerged as attractive alternatives in the demand for clean, renewable energy sources. However, the CST field requires the improvement of its competitiveness in comparison with other more mature technologies by the increase of conversion efficiency (solar to electricity), while reducing investment costs (IEA 2014).

The maximum temperature of the working cycle determines the power that can be extracted from the thermal energy and the efficiency increases with higher temperatures. Therefore, higher operating temperatures enhance turbine efficiencies and also reduce storage volumes because the higher the temperature of the HTF in the receiver increases, the less storage material for the same storage capacity is needed (the storage capacity is given by the product of the mass of storage material and the enthalpy difference at the low and high temperatures).

The efficiency of the receivers depends on the technology, while the maximum possible efficiency of the conversion of heat into electricity is given by Carnot theorem which determines the ideal thermodynamic efficiency. Conversely, the overall efficiency of the STE plant is the product of the receiver efficiency by the Carnot efficiency and a fixed coefficient which defines the imperfection of the thermodynamic engine.

M. I. Roldán Serrano (✉) · J. F. Reche · E. Zarza Moya
CIEMAT-Plataforma Solar de Almería, Ctra. de Senés, km. 4.5, E04200, Tabernas, Almería, Spain
e-mail: mabel_roldan@yahoo.es

J. F. Reche
e-mail: jesus.fernandez@psa.es

E. Zarza Moya
e-mail: eduardo.zarza@psa.es

Z. Driss et al. (eds.), *CFD Techniques and Thermo-Mechanics Applications*,
https://doi.org/10.1007/978-3-319-70945-1_2

Point-focus systems can convert a larger fraction of the energy absorbed by the receiver into electricity than linear systems. Thus, central receiver systems (CRS) are potentially less costly in the long term than other line-focus technologies because they are able to achieve higher temperatures and utilise more efficient thermodynamic cycles (IEA 2014).

In CRS, the solar radiation is concentrated by a field of heliostats onto the receiver. For a tubular design of a conventional CRS, the incident energy reaches the tube walls which transfer the heat to a fluid, the so-called heat transfer fluid. The heat transfer fluid (HTF) passes directly through a heat exchanger to generate steam for a Rankine cycle. The system can integrate thermal storage, either combined into the solar heating loop or through the use of a third loop with an additional heat exchanger (Ortega et al. 2008).

Therefore, the selection of an appropriate HTF influences the cost of the solar receiver, thermal storage and heat exchanger, and determines the achievement of high receiver and cycle efficiencies. The desirable properties for a HTF are (Becker 1980):

- Good heat exchange and high thermal conductivity for cost-efficient receivers and heat exchangers;
- High heat capacity to reduce the storage volume and mass flows required;
- Low viscosity to reduce pressure drop and pumping losses;
- Stability at high temperatures and a high boiling point to allow for low overpressure requirements and avoid HTF degradation. A high upper temperature limit also leads to efficient thermodynamic cycles;
- Low melting point to minimise the use of heat tracing in the facility;
- Low cost;
- Low vapour pressure (<1 atm) at high temperature and
- Low corrosion with metal alloys used to contain the HTF.

It is important to consider that there is not an ideal HTF and the selection depends on the facility requirements. Additionally, other aspects should be taken into account, such as (Boerema et al. 2012):

- Fluid availability;
- Fluid and plant life expectancy;
- Environmental issues;
- Operation and maintenance (safety in plant operation);
- Additional components requirements (expansion vessels, heat tracing, etc.) and
- Compatibility with other materials.

One of the main HTFs used in CRS is a nitrate salt mixture with a melting point of about 240–250 °C and an upper temperature limit of around 585 °C. At temperatures of about 600 °C, this fluid becomes thermally and chemically unstable because nitrogen and oxygen decompose to form gases such as O_2, N_2 and several nitrogen oxides depending on environment conditions (Kearney et al. 2003).

Other commercial plants use water/steam as working fluid at both the solar receiver and the power block which simplifies the CRS because it does not require an intermediate heat exchanger located between the receiver and the power block. This fact leads to improved efficiencies and also a decrease in electricity production costs. Nevertheless, one of the most important problems of water/steam is its scarcity in desert regions, because CST plants are considered a very promising option for these zones, where large land area and high direct solar radiation are available. Other issues associated with the use of water/steam are: high-temperature steam corrosion in several different alloys (Vignarooban et al. 2015), the high operating pressure required (≥ 115 bar) and complex controls due to the two-phase flow in the receiver.

Thus, alternative HTFs are being studied to utilise them in CRS. The analysis of systems with fluid flows, heat transfer and other associated phenomena, such as chemical reactions, can be addressed by means of computer simulation and proper numerical algorithms. The field of Fluid Mechanics concerned with this analysis is called computational fluid dynamics (CFD).

CFD provides a reliable tool to assist engineers and researchers in the design and analysis of industrial equipment or innovative facilities often reducing or eliminating the need for performing trial-and-error experimentation. Therefore, CFD simulation provides rapid information of the physical phenomena involved in the problem analysed as well as a reduction of the investment cost in the research developed (Al-Baghdadi 2014).

In this context, an innovative HTF for CRS should be proposed and studied in order to improve both the receiver efficiency and the power cycle performance. This chapter describes the main innovative HTFs studied in CRS so far and proposes the evaluation of an alternative HTF (supercritical fluid) in comparison with a conventional one (molten salt) using CFD simulation.

2 Innovative Heat Transfer Fluids in Central Receiver Systems

As mentioned, the HTF is a key component of CST plants as it has a significant influence on the receiver efficiency, defines the type of power cycle and the performance achieved and determines the thermal storage technology used. HTFs can be divided into several different groups: liquid, gas, two-phase, particulate and supercritical fluids (Benoit et al. 2016). In this section, the main characteristics of innovative fluids belonging to each group are summarised.

2.1 Liquid Metals

Liquid metals have been used in nuclear industries since 1940s and are being studied for use in CST systems as an HTF and thermal energy storage medium. They have not been utilised in commercial CST applications yet, but they are efficient HTFs that could lead to the further development of CST systems.

Their main promising properties are extensive operating temperature range, low viscosity and higher heat transfer coefficients than those of other liquid fluids. As reference, liquid sodium has an operating temperature range between 98 and 883 °C (Vignarooban et al. 2015).

These materials are relatively costlier than molten salts or water/steam, they have strong corrosion properties for current materials, and their low specific heat capacity in comparison with nitrate/nitrite based salts makes them less favourable to be used as thermal storage media (Benoit et al. 2016).

2.2 Innovative Molten Salts

Advanced molten salts have been studied and developed for solving their main limitations: high melting point which leads to freezing issues and thermal instability that changes the HTF properties. Thus, a mixture of nitrate salts of lithium, sodium, potassium, caesium and calcium exploits the eutectic behaviour resulting in a melting point of 65 °C (lower than the 142 °C of the conventional Hitec salt) and a thermal stability up to over 500 °C (same range of conventional nitrate salts). The melting point reduction is mainly due to caesium addition, which forms the quaternary eutectic. However, the cost of raw materials for this advanced fluid is higher than the conventional binary salts. Such a high cost can be minimised by optimising the mixture to limit caesium and lithium nitrate amounts while maintaining acceptable physical properties (Benoit et al. 2016; Vignarooban et al. 2015).

Halide-based molten salts reach working temperatures of 800 °C and they consist of five species of halide salts. This fluid is still under development and initial results about corrosion show that nickel-based alloys can have a good resistance. The increase of the equipment lifetime would reduce maintenance costs because these alloys are four times more expensive than iron-based steels with a lower corrosion resistance (Benoit et al. 2016).

Additionally, new studies about the use of nanoparticles in molten salts are being developed. Dispersing solid particles in fluids enhances thermal conductivity, but these solids rapidly settle in fluids. Nanoparticles improved with surfactants and stabilisers could remain in suspension and have a million time larger surface area per unit volume than the contact surface of microparticles, allowing improved heat transfer properties (IEA 2014).

2.3 Innovative Gas Fluids

Atmospheric air has been proposed as an alternative HTF in solar CRS because of its wide operating temperature range, low cost, abundance in nature, ease of use and environmental friendliness. In contrast, it presents several disadvantages, such as limited heat transfer properties and large pumping power. In addition, corrosion issues have also been found in carbon steels due to their contact with air at high temperatures. The very low dynamic viscosity of the air compared to the one of molten salts or liquid metals gives good flow properties inside the pipelines that favour the heat transfer and may compensate its low thermal conductivity (Tian and Zhao 2013).

Other gases such as helium have also been investigated for use in CST plants. Helium was widely utilised as a working substance in high-temperature nuclear re-actors. It is also relatively affordable, since it can be generated in natural gas extraction processes. Helium can also be used at very high temperatures (similar to air), it is chemically inert, and its specific heat is five folds that of the air. However, the major drawbacks are the low heat capacity and heat transfer between the fluid and the internal surfaces of the pipelines. Consequently, high pressures and high fluid velocities are required (Benoit et al. 2016; Massidda and Varone 2007).

Hydrogen has been used as HTF in solar dish–Stirling systems and presented as a promising energy carrier and might be the best answer for energy storage. It is the most abundant element in nature and the lightest weight existing gas, but the main drawback is that it should be obtained from other elements, which implies a significant drain of energy and, as a consequence, an increased cost (Benoit et al. 2016).

In general, the use of gases as HTF in CST plants requires an optimisation of the operating strategy together with the periodic maintenance of the facility to avoid leakage problems. These issues are minimised in central receiver systems because they have a fixed receiver and a shorter piping line in comparison with linear-focus receivers (Roldán and Fernández-Reche 2016).

2.4 Particle Suspensions

Particle receivers consider a variety of particle circulation concepts: moving bed, falling film, bubbling fluidised bed, spouted beds, upflow bed and circulating fluidised beds. Direct solar absorption in particles does not require a quartz window and accepts very high solar flux density (around 1 MW/m^2). Nevertheless, particle flow stability is difficult to control and may present high convection losses. On the other hand, indirect absorption receivers allow for the operation with lower solar flux density (200–400 kW/m^2), but offer a better control of particle circulation and the management of both operating pressure and atmosphere composition is possible (Tan and Chen 2010; Benoit et al. 2016).

Furthermore, solid particles have an inherent storage capacity which enables their use as HTF. In fact, sand-like ceramic particles may achieve temperatures of around 1000 °C. They can be stored in an insulated tank and/or used to heat a secondary working fluid for the power cycle.

Some desirable properties for the solid particles are (Ho and Iverson 2014):

- High packing density;
- High heat capacity;
- Resistance to mechanical and thermal shocks;
- Resistance to sintering and agglomeration;
- Corrosion resistance in air and other media;
- High solar absorptance and low thermal emittance and
- Low cost and wide availability.

Computational fluid dynamics models of falling particle receivers have been developed in order to predict the performance of this alternative receiver concept (Tan and Chen 2010).

2.5 *Supercritical Fluids*

Supercritical fluids can be operated at very high temperatures and are able to act as both HTF for solar receivers and working fluid for the power block. The increasing interest in s-CO_2 Brayton cycles, able to reach thermodynamic efficiencies above 50% at concentration ratios and temperatures achievable by concentrated solar radiation, has led to propose s-CO_2 as HTF in CST systems (Ho and Iverson 2014). Therefore, s-CO_2 can achieve higher conversion efficiencies at lower temperatures in comparison with steam Rankine and helium cycles.

However, due to the high pressure drop at the required operating conditions, s-CO_2 is not appropriate for parabolic trough plants because there are extensive pipelines used. Thus, the most compatible CST technology is the central receiver system, considering fluid temperatures above 750 °C (Vignarooban et al. 2015).

Tubular receivers that employ s-CO_2 as the heat transfer fluid are a possible design as the small diameter tubes may enable the high pressures required to achieve high power block efficiency. The pressures expected at the turbine inlet for s-CO_2 Brayton cycle are around 15–25 MPa. The integration of s-CO_2 with storage is not a viable option because of the high operating temperatures required. Hence, in this case, the storage system should be separated with an intermediate heat exchange considering other storage media (Ho and Iverson 2014).

Because of the increasing interest in using supercritical fluids as heat carriers, this chapter aims to evaluate the thermal behaviour of s-CO_2 in a tubular receiver for a solar tower by determining the operating conditions required to maintain the supercritical phase, and also to obtain an initial optimisation of the receiver design selected. To this end, a CFD model has been developed to compare the s-CO_2 behaviour with the one of a conventional fluid (molten salt), considering the same receiver design. From the initial results, the optimisation of the tubular receiver design will be analysed.

3 Supercritical Fluids

3.1 Advantages and Disadvantages

Supercritical fluids are substances for which both temperature and pressure are above the critical values, where distinct liquid and gas do not exist (Fig. 1). These fluids combine gas-like viscosity and diffusivity, and liquid-like density and solvating properties, which makes them excellent solvents for various applications (Knez et al. 2014).

In general, processes involving supercritical fluids are sustainable, environmentally friendly and cost-efficient, due to the possibility of separating and drying the resulting product by simple expansion. Furthermore, this type of fluids can be recovered, recycled and reused without considering any intermediate purification step. Industrial processes with supercritical fluids present environmental advantages such as low energy consumption during the operation and its ability to replace more hazardous conventional organic solvents.

Thus, the main positive aspects of supercritical substances are (Knez et al. 2014):

- Thermodynamic stability;
- Excellent heat transfer properties;
- Modification of thermophysical properties by controlling operating conditions (pressure and/or temperature);
- Excellent solvent for various applications and
- Environmentally friendly (supercritical water and CO_2).

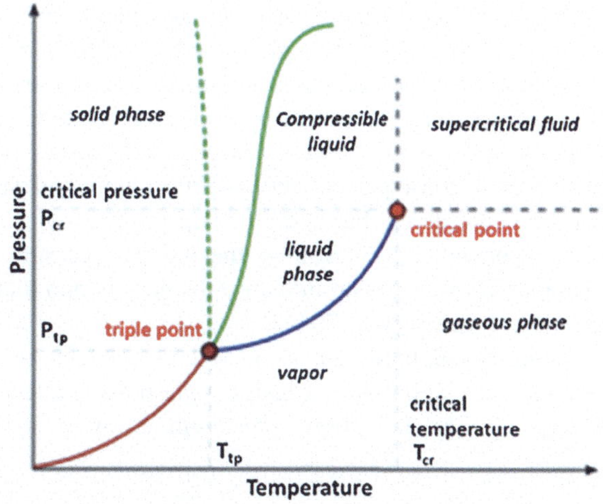

Fig. 1 Pressure–temperature phase diagram

Due to their good heat transfer properties, supercritical fluids are considered as alternative working fluids in power cycles, and also as environmentally benign heat transfer fluids.

Nevertheless, these fluids present some drawbacks such as:

- High operating pressures are required to maintain the supercritical conditions;
- Leakage issues in facilities with moving connections, due to their relative low density (better than gas density and lower than liquid one) and their high operating pressures;
- Operating conditions for supercritical fluids (high temperature and pressure) involve specific equipment and safety demands in the facility;
- The direct thermal storage of supercritical fluids is not a viable option (use of an intermediate heat exchanger with a separate storage media) and
- Some industrial applications with supercritical fluids may be often costlier than the conventional methods because of the technical requirements.

As a consequence, a compromise should be found depending on the application selected and the operating conditions required to optimise the use of supercritical fluid.

3.2 Applications

Nowadays, some technologies based on the utilisation of supercritical fluids have emerged because of their environmental and energetic advantages compared to the conventional processes. The good solubility of many natural compounds (vitamins, aromas, natural pigments or essential oils) in supercritical fluids is used to extract valuable products such as food additives and antioxidants. Conventional extraction methods involve organic solvents, which are in contact with the raw material and dissolve the final product. The solution with the valuable product is treated by high-temperature methods in order to separate the solvent which requires some purification steps to be reused. The advantages of using supercritical fluids for the generation of natural products are related to the possibility of removing the solvent from products by simple depressurisation. Another positive aspect is the better stability of the natural compound obtained because of the lower process temperatures. As a consequence, the original compound structure without the organic solvent is preserved. Furthermore, supercritical fluids allow the selective extraction of components by controlling the applied pressure and temperature which are operating conditions that influence the density and dielectric constant of the natural compound (Knez et al. 2014).

Supercritical fluids are also used in polymer processing because the fluid is absorbed by the polymer, causing a plasticising effect that reduces the viscosity of the system. This fact allows processing at lower temperatures.

Another application is the supercritical drying and cleaning to remove liquids from solids without altering the structure of the solid. This process is based on the liquid-like supercritical properties that enable the capacity of dissolving organic solvents.

Furthermore, supercritical fluids are used as new reaction media for chemical and biochemical processes (alkylation, polymerisation, oxidation, metathesis, hydrogenation and ammonolysis, among others). They are also applied in synthesis of new materials and new catalyst supports such as aerogels, and in special separation techniques such as chromatography.

The most important utilisation of supercritical fluids for CST technologies is the one devoted to transfer the heat in refrigeration systems and power cycles. The excellent heat transfer and thermodynamic properties of these fluids found application in the field of energy production. Conventional power cycles currently used in STE plants involve either water/steam or molten nitrate salt, as heat transfer and/or working fluid, and have a thermal-to-electricity conversion efficiency between 30 and 40% (Ho and Iverson 2014). As mentioned, there is a research field whose objective is the search for heat transfer fluids that can withstand high temperatures and enable system designs with low radiative and convective heat losses, together with high thermal and thermal-to-electricity efficiencies.

Thus, s-CO_2 can be used as a working fluid in both Rankine cycle solar systems and Brayton cycle concepts. The main advantage of the supercritical Rankine cycle is that, during heating, the working fluid does not pass through a distinct two-phase region, obtaining higher thermal efficiency. Supercritical Brayton cycles involve direct heating of s-CO_2 in tubular receiver able to withstand high pressures and temperatures, and they are considered in CST technologies due to their greater efficiency (50%).

Therefore, appropriate designs for tubular central receivers are required in order to enhance the overall efficiency of the CST plant by using s-CO_2. Next section describes the developed CFD analysis to compare the thermal behaviour of s-CO_2 (innovative fluid) with that of molten salts (conventional fluid) considering an initial tubular design in order to optimise it afterwards for the supercritical fluid.

4 CFD Analysis of a Supercritical Fluid Used as Heat Transfer Fluid in a Solar Tower Tubular Receiver

This analysis has been developed because emerging receiver designs consider the achievement of higher efficiencies using advanced power cycles, such as s-CO_2 closed-loop Brayton cycles. For this purpose, the s-CO_2 direct heating is considered in tubular receiver designs capable of withstanding high internal fluid pressures (around 20 MPa) and temperatures (900 K) (Ho and Iverson 2014). Due to the high pressures required and the presence of moving components installed in the pipelines (ball-joints and/or flexible connections), the use of s-CO_2 presents many technical challenges caused by the compatibility of seal materials and fluid leakages of the moving connections. These problems can be better controlled in solar tower systems because the receiver is fixed (Vignarooban et al. 2015).

In this regard, a preliminary CFD analysis has been developed using the design of a molten-salt tubular receiver which was previously tested. In order to minimise the computational cost, a simplified geometry has been considered in the model. CFD simulations have been carried out to determine the minimum operating conditions required for the use of s-CO_2 as an HTF and to compare the efficiencies reached by both fluids. From this information, an initial optimisation of the original design has been developed.

4.1 Description of the Initial Tubular Receiver Design

The selected receiver consists of three different modules through which the HTF flows consecutively. Each module is made of 20 alloy tubes that receive the solar radiation coming from the heliostat field and transfer the heat to the working fluid. Figure 2a shows the fluid flowing through the central module before it passes through the lateral ones. There is a manifold at the inlet and outlet of each module to homogenise the flow conditions.

As shown, the simulation domain considers the symmetrical behaviour of the fluid in the module and, for the evaluation of the minimum operating conditions required to maintain the supercritical phase, the central module where the fluid inlet is located was selected.

For a further optimisation of the complete receiver design, the solution domain selected has been considered for the three modules in order to obtain the overall operating conditions of the tubular receiver.

4.2 Supercritical Fluid CFD Analysis Procedure

In order to develop a preliminary study of the s-CO_2 behaviour in comparison with that of molten salt, a 2D symmetrical simulation domain of module 2 (Fig. 2b) has been considered. This initial approach was regarded to minimise the computational cost and obtain initial results which would lead the study towards a further optimisation of the tubular receiver design.

The CFD model created was validated by experimental data obtained from tests of the molten-salt receiver, taking into account the measurements at a quasi-steady state selected. The measured irradiance distribution was considered as boundary condition and the pressure and mass flow at the inlet were those measured at the pump impulsion. The fluid temperature was measured at the inlet and outlet of each module and the pressure was used as boundary condition in order to compare the numerical fluid temperature with the experimental one.

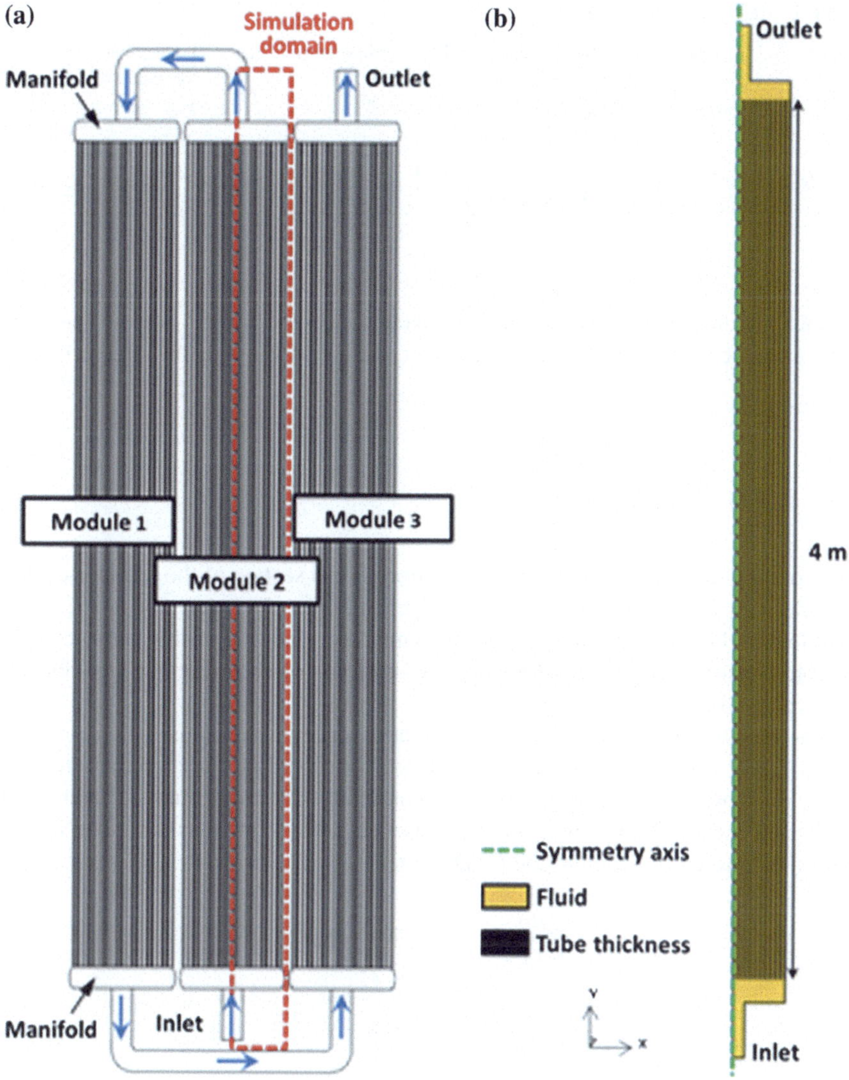

Fig. 2 Geometry: **a** scheme of the tubular receiver, **b** simulation domain (Roldán and Fernández-Reche 2016)

Furthermore, simulations with s-CO_2 as an HTF were developed at different fluid inlet temperatures (800, 715, 600 and 500 K) to study the operating conditions required for each case. Table 1 summarises the steady-state conditions for each fluid.

In addition, an initial optimisation of the receiver design has been developed considering a greater length of the tubes (10 m) and doubling the number of tubes

Table 1 Steady-state conditions

Fluid	Inlet pressure (Pa)	Outlet pressure (Pa)	Fluid temperature at the inlet (K)	Total power received (kW)
Molten salts	589,000	347,000	715	720
s-CO_2	11,000,000	7,500,000	800, 715, 600, 500	720

Source Roldán and Fernández-Reche (2016)

in each module. The new design has been conceived taking into account the comparison between the thermal behaviour of the s-CO_2 and that of the molten salts.

4.3 CFD Modelling

The solution domain selected was a two-dimensional section of the receiver central module with the following zones: inlet pipe, inlet manifold, receiver tubes, outlet manifold and outlet pipe. Regarding the symmetry condition of this geometry, the domain consisted of two sub-domains: the thickness of each receiver tube and the fluid (Fig. 2b).

The model is based on the equations that describe the dynamic behaviour of a fluid. These equations come from the three conservation laws: conservation of mass, momentum and energy. Thus, the continuity (1), momentum (2) (Bachelor 1967) and energy (3) (Versteeg and Malalasekera 1995) equations described for the fluid medium are included in the CFD model:

$$\frac{\partial \rho}{\partial t} + \nabla \cdot (\rho \vec{v}) = S_m \quad (1)$$

$$\frac{\partial}{\partial t}(\rho \vec{v}) + \nabla \cdot (\rho \vec{v} \vec{v}) = -\nabla p + \nabla \cdot (\bar{\bar{\tau}}) + \rho \vec{g} + \vec{F} \quad (2)$$

$$\frac{\partial \rho}{\partial t}(\rho E) + \nabla \cdot (\vec{v}(\rho E + p)) = \nabla \cdot \left(k_{\mathrm{eff}} \nabla T - \sum_j h_j \vec{J}_j + (\bar{\bar{\tau}}_{eff} \cdot \vec{v}) \right) + S_{\mathrm{he}} \quad (3)$$

where ρ is the density of the fluid, t is elapsed time, $\vec{v}$ is the velocity vector with respect to the coordinate system, S_{m} is the mass source, p is the static pressure, $\bar{\bar{\tau}}$ is the stress tensor, $\rho \vec{g}$ is the gravitational body force, $\vec{F}$ is the external body force, E is the energy transfer $\left(E = h - \frac{p}{\rho} + \frac{v^2}{2}\right) k_{\mathrm{eff}}$, is the effective conductivity which includes the turbulence thermal conductivity, h_j is the enthalpy of species j, $\vec{J}_j$ is the diffusion flux of species j, $\bar{\bar{\tau}}_{eff}$ is the viscous stress tensor, and S_{he} is the volumetric heat source. Therefore, the proposed CFD model requires solving the continuity, momentum and energy equations adapted to the steady-state flow condition.

The selected mesh is made up of quadrilateral cells because the structured grid favours convergence. In order to select the best mesh size, it is recommended to develop a mesh independence test to adjust the grid size to the minimum one that ensures an independent solution of the cell number. Furthermore, the mesh quality has been evaluated by the equiangle skew (Q_{EAS}) parameter. The 100% of the cells are in the Q_{EAS} range 0–0.44 which corresponds to both the excellent quality range (0–0.25) and the good one (0.25–0.5). It means that the mesh quality is appropriate because all the grid elements are within the good quality range. Figure 3b shows the adaptation of the mesh selected to the geometry.

The material properties considered were obtained from literature and the boundary conditions selected were: inlet/outlet pressure, natural convection in outer walls because insulation does not eliminate heat losses to the receiver cavity, symmetry condition for the cross-cutting axis of the geometry, and the coupled condition for the contact walls (Fig. 3b).

Furthermore, the volumetric heat source for each absorber tube was obtained from the analysis of the heat-flux distribution measured (Fig. 3a), showing an average irradiance of 122 kW/m^2. In order to define the tube zone with the maximum heat flux, each tube length of the receiver was divided into three parts according to the heat-flux distribution. The volumetric heat source was

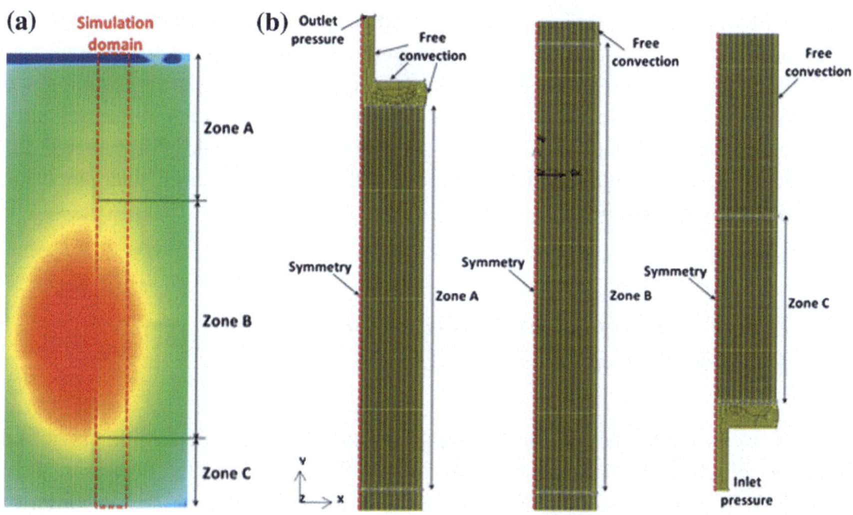

Fig. 3 Definition of the boundary conditions: **a** heat-flux distribution, **b** boundary conditions in the solution domain (Roldán and Fernández-Reche 2016)

Table 2 Comparison between experimental and numerical data for model validation

Experimental inlet pressure (Pa)	Numerical inlet pressure (Pa)	Deviation (%)
589,000	550,023	6.62

Source Roldán and Fernández-Reche (2016)

implemented in each zone depending on its volume (404,651, 1,540,465 and 845,581 W/m^3 for zones A, B and C, respectively).

Both the definition of a 2D symmetrical geometry and the approach of a constant heat source for each tube zone reduce the computational cost. Moreover, the boundary conditions were adapted to the HTF used. The outlet pressure was set at 0.35 MPa for molten salts (measured value) and 7.50 MPa for s-CO_2 (critical pressure of 7.38 MPa) in order to study the initial operating conditions that ensure the supercritical phase.

The viscosity model has been selected according to the Reynolds number (Re) evaluated for molten salts in three zones: receiver tube, manifold and module inlet/outlet. The Re value was between 6.12×10^5 and 9×10^6 which belongs to a turbulent regime. Nevertheless, κ-ε renormalisation group (RNG) model was used to take into account with low Re areas.

4.4 Validation

The CFD model was validated taking into account a molten-salt test. The variables considered were the absolute inlet pressure and the outlet temperature of the fluid at the steady state selected. In this case, the outlet pressure was fixed as boundary condition. The deviation between experimental and numerical data is summarised in Table 2. According to previous studies, high-temperature solar processes simulation accepted deviations of around 9% (Ozalp and Jayakrishna 2010), the obtained numerical results are in agreement with the experimental data since the deviation is lower than 7%.

4.5 Initial Operating Conditions of the Tubular Receiver for s-CO_2

In order to evaluate the s-CO_2 behaviour as an HTF and determine its initial operating conditions, simulations with different s-CO_2 inlet temperatures were developed. The outlet pressure was set at 7.5 MPa in a first approach in order to ensure the supercritical condition because the subsequent optimisation study takes into account the three receiver modules. Thus, the required inlet pressure was around 9.67 ± 0.05 Pa, and the temperature reached by the fluid is included in Table 3.

Table 3 Simulation results for s-CO_2

Simulation	Inlet temperature (K)	Outlet temperature (K)	Heat gained by the fluid (kW)	Maximum temperature reached (K)
1	800	812	358.27	1038
2	715	726	321.18	903
3	600	608	226.49	758
4	500	506	167.46	607

Source Roldán and Fernández-Reche (2016)

The largest increase in heat gained by the fluid was obtained at the inlet temperature of 800 K. However, further research has been proposed to evaluate the radiation heat losses in addition to the convection losses considered for this case. Thus, simulation data for 715 K were considered in a more conservative approach to analyse the initial operating conditions required for the s-CO_2. Figure 4 shows a non-homogeneous distribution for both of fluid temperature (Fig. 4a) and pressure (Fig. 4b). The receiver design, then, needs to be adapted to the supercritical fluid in order to obtain more homogeneous operating conditions. Hence, the optimisation of the tubular receiver design is proposed to improve the fluid dynamic behaviour in the receiver.

Simulation results obtained from the use of s-CO_2 as HTF, considering the inlet temperature of 715 K, are summarised in Table 4.

The operating pressure for s-CO_2 is much higher than the one required by the molten salts (Table 2). Nevertheless, the pressure required is achievable by

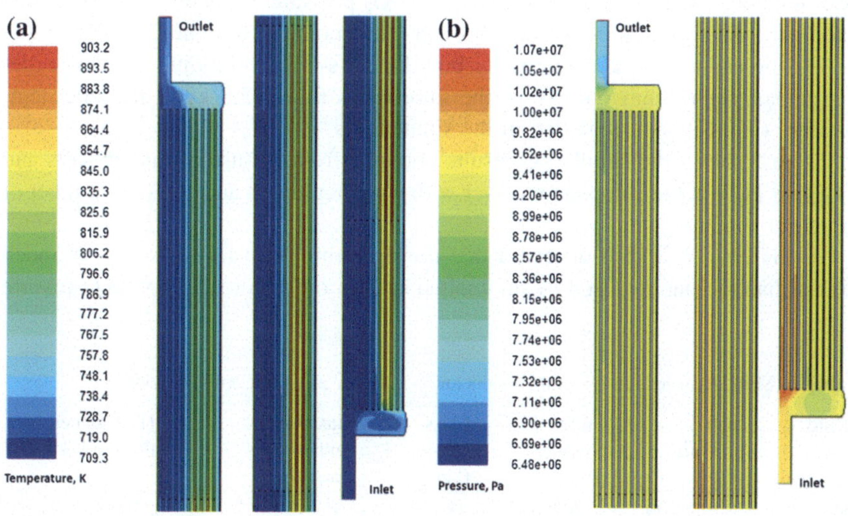

Fig. 4 Simulation results: **a** thermal distribution of s-CO_2 (T_{inlet} = 715 K), **b** pressure distribution of s-CO_2 (T_{inlet} = 715 K) (Roldán and Fernández-Reche 2016)

Table 4 Simulation results for s-CO_2

Fluid	Inlet pressure (Pa)	Outlet pressure (Pa)	Mass flow (kg/s)	Maximum temperature reached (K)	Heat gained by the fluid (kW)
s-CO_2	9,695,240	7,500,000	24.90	903	321.18

Source Roldán and Fernández (2016)

equipment used in other solar facilities (Muñoz-Anton et al. 2014). Furthermore, the maximum temperature reached is higher than the upper working temperature limit for molten salts (873 K). From these results, s-CO_2 seems to be a promising alternative as HTF in tubular central receivers, but an initial optimisation of the selected design considering more homogeneous conditions for both fluids has to be developed to obtain a more standardised evaluation criterion.

4.6 Optimisation of the Receiver Design for s-CO_2

In order to evaluate the complete receiver and study an initial optimisation of its design, new CFD simulations have been developed considering a similar domain to the previous one with a greater tube length (10 m) and 40 tubes for each module. These CFD simulations were based on the previous ones developed for the receiver central module, taking into account the symmetry condition. However, in order to evaluate the three receiver modules, the outlet conditions of the simulated module were considered as the inlet conditions of the following module which is going to be simulated. As reference condition, the s-CO_2 pressure at the receiver outlet was fixed according to the inlet power Brayton cycle pressure (around 24 MPa).

This procedure was considered for both fluids (s-CO_2 and molten salts) in order to compare their behaviour. Thus, the same mass flow (95.3 kg/s) and inlet temperature (750 K) were regarded in the simulations.

Table 5 shows the simulation results obtained from the use of molten salts and s-CO_2 as HTF in the complete receiver with a greater length and a higher number of tubes per module.

In this case, it is also observed that the operating pressure for s-CO_2 is much higher than the one required by the molten salts in order to reach the inlet pressure

Table 5 Simulation results for molten salts and s-CO_2 for an initial optimisation

Fluid	Inlet pressure (Pa)	Outlet pressure (Pa)	Mass flow (kg/s)	Maximum temperature reached (K)	Heat gained by the fluid (kW)
Molten salts	551,709	550,973	95.3	895	22060.8
s-CO_2	24,216,635	24,201,324	95.3	963	25216.6

for the power cycle selected (Brayton). Furthermore, the heat gained by s-CO_2 at the outlet of the receiver module is around 12% greater than the one obtained by molten salts, considering the same mass flow for both fluids. Hence, it is still demonstrated that using the s-CO_2 as HTF in solar tower receivers can be a promising option, but it is necessary to take into account the high operating conditions required for its use in the solar facility which might increase the pumping consumption and enhance corrosion issues.

4.7 Conclusions Obtained from the CFD Analysis

A simplified CFD model has been developed in order to obtain a preliminary analysis of a tubular solar receiver considering s-CO_2 as HTF. Thus, the simulation domain selected consisted of a 2D symmetrical geometry of the central receiver module. The model validation was carried out by the comparison between experimental data coming from a tested molten-salt tubular receiver and numerical results, regarding the operating conditions at a selected steady state.

The validation took into account measurements of the fluid pressure at the inlet and outlet of the central receiver module (simulation domain). The comparison between these data showed a maximum deviation lower than 7%. Hence, the CFD model was used to study the s-CO_2 behaviour for several different inlet temperatures.

The optimum inlet temperature of s-CO_2 was set to 715 K to determine the minimum operating conditions to maintain the supercritical phase. The operating pressure for s-CO_2 is much higher than the one required by the molten salts, but it is achievable by the equipment used in solar facilities for parabolic trough collectors. Furthermore, the maximum temperature reached is higher than the upper working temperature limit for molten salts (873 K).

From these results, s-CO_2 seems to be a possible alternative as HTF in tubular central receivers, but an initial optimisation of the selected design considering both fluids has been developed to obtain further information.

The CFD analysis of a new geometry with a greater tube length and a higher number of tubes, considering the same mass flow and inlet temperature for both fluids, has led to conclude that the heat gained by s-CO_2 was around 12% greater than the one captured by molten salts. Thus, the use of s-CO_2 as HTF in solar tower receivers appears to be a promising alternative, taking into account both the operating conditions required and their maintenance cost.

5 Summary and Conclusions

The selection of an appropriate HTF is important for increasing both the efficiency of the solar receiver and that of the overall STE plant. In addition, the relative cost of a solar receiver can be minimised by the selection of a HTF capable of achieving higher temperatures or, as a result, higher receiver efficiencies. Higher working fluid temperature is associated with greater thermal efficiency in both receiver and power cycle if a suitable HTF and receiver design are defined.

Current commercial HTFs present some limitations that do not allow the improvement of thermal efficiencies. These limitations are due to the fluid degradation and its thermal instability from temperatures of about 600 °C. Therefore, in order to improve the thermal-to-electricity efficiency of the facility, the integration of more efficient power cycles working at higher temperatures in the CST facility is required. To this end, alternative HTFs must be studied for application in CST technologies.

In this context, new supercritical s-CO_2 power cycles have been proposed, and the utilisation of this fluid as HTF at high operating pressures enables both its direct use in the power block with a possible hybridisation system or a thermal storage with an intermediate storage medium.

In order to study the feasibility of s-CO_2 as HTF in central receivers, this chapter describes the CFD evaluation of a tubular receiver design in order to define the operating conditions required and its thermal behaviour in comparison with a conventional fluid (molten salts). Simulations results show that the use of s-CO_2 as HTF in solar tower receivers appears to be a promising alternative, but other operation aspects such as pumping consumption and possible corrosion effects should be taken into account in further assessments.

Acknowledgements This work is partially supported by both ALCCONES Project (S2013/MAE-2985) through the Program of R&D activities between research groups in Technologies 2013, co-financed by structural funds, and the Spanish government in the framework of the DETECSOL project (Ref. ENE2014-56079-R) with ERDF funds.

The authors also gratefully acknowledge the helpful comments and suggestions of the reviewers, which have improved the presentation.

References

Al-Baghdadi MARS (2014) Computational fluid dynamics applications in green design. International Energy and Environment Foundation, Iraq

Bachelor GK (1967) An introduction to fluid dynamics. Cambridge University Press, Cambridge

Becker M (1980) Comparison of heat transfer fluids for use in solar thermal power stations. Electr Pow Syst Res 3:139–150

Benoit H, Spreafico L, Gauthier D, Flamant G (2016) Review of heat transfer fluids in tube-receivers used in concentrating solar thermal systems: Properties and heat transfer coefficients. Renew Sustain Energ Rev 55:298–315. https://doi.org/10.1016/j.resr.2015.10.059

Boerema N, Morrison G, Taylor R, Rosengarten G (2012) Liquid sodium versus Hitec as heat transfer fluid in solar thermal central receiver systems. Sol Energy 86:2293–2305. https://doi.org/10.1016/j.solener.2012.05.001

Ho CK, Iverson BD (2014) Review of high-temperature central receiver designs for concentrating solar power. Renew Sust Energ Rev 29:835–846. https://doi.org/10.1016/j.rser.2013.08.099

International Energy Agency (IEA) (2014) Technology roadmap: solar thermal electricity. https://www.iea.org/publications/freepublications/publication/technologyroadmapsolarthermalelectricity_2014edition.pdf. Accessed 20 May 2016

Kearney D, Herrmann U, Nava P, Kelly B, Mahoney R, Pacheco J, Cable R, Potrovitza N, Blake D, Price H (2003) Assesment of a molten salt heat transfer fluid in a parabolic trough solar field. J Sol Energy Eng 125(2):170–176. https://doi.org/10.1115/1.1565087

Knez Ž, Markočič E, Leitgeb M, Primožič M, Knez Hrnčič M, Škerget M (2014) Industrial applications of supercritical fluids. Rev Energy 77:235–243. https://doi.org/10.1016/j.energy.2014.07.044

Massidda L, Varone A (2007) A numerical analysis of a high temperature solar collecting tube using helium as a heat transfer fluid. CRS4 report

Muñoz-Anton J, Biencinto M, Zarza E, Díez LE (2014) Theoretical basis and experimental facility for parabolic trough collectors at high temperature using gas as heat transfer fluid. Appl Energy 135:373–381. https://doi.org/10.1016/j.apenergy.2014.08.099

Ortega JI, Burgaleta JI, Téllez F (2008) Central receiver system solar power plant using molten salt as heat transfer fluid. J Sol Energy Eng 130(2):024501. https://doi.org/10.1115/1.2807210

Ozalp N, JayaKrishna D (2010) CFD analysis on the influence of helical carving in avortex flow solar reactor. Int J Hydrogen Energy 35:6248–6260. https://doi.org/10.1016/j.ijhydene.2010.03.100

Roldán MI, Fernández-Reche J (2016) CFD analysis of supercritical CO_2 used as HTF in a solar tower receiver. AIP Conf Proc 1734:030031. https://doi.org/10.1063/1.4949083

Tan TD, Chen YT (2010) Review of study on solid particle solar receivers. Renew Sustain Energy Rev 14(1):265–276. https://doi.org/10.1016/j.rser.2009.05.012

Tian Y, Zhao CY (2013) A review of solar collectors and thermal energy storage in solar thermal applications. Appl Energy 104:538–553. https://doi.org/10.1016/j.apenergy.2012.11.051

Versteeg HK, Malalasekera W (1995) An introduction to computational fluid dynamics: The finite volume method. Longman Scientific & Technical, Harlow

Vignarooban K, Xu X, Arvay A, Hsu K, Kannan AM (2015) Heat transfer fluids for concentrating solar power systems—A review. Appl Energ 146:383–396. https://doi.org/10.1016/j.apenergy.2015.01.125

Computational Fluid Dynamics for Thermal Evaluation of Earth-to-Air Heat Exchanger for Different Climates of Mexico

M. Rodríguez-Vázquez, I. Hernández-Pérez, J. Xamán, Y. Chávez and F. Noh-Pat

Currently, the energy requirements that our society uses to carry out its activities, are increasing with citizens' wealth. In order to meet these requirements, power plants will have to increase the use of fossil fuels, resulting in an environmental pressure due to the greenhouse gases emission into the atmosphere. In Mexico, 88% of the energy is produced by burning hydrocarbons and 95% of the total energy produced is intended to meet the social and economic needs for energy (national energy balance, 2014). From this 95%, a 27% is consumed in by the residential sector, making it the third largest consumer of electricity. Mexico is not a unique case, which has caused a good number from the scientific community to seek alternatives to reduce energy consumption and focus on the study of passive solar systems. Passive solar systems are used to heat, cool, and illuminate buildings and do not require a mechanical device to function. Some examples of these systems are the

M. Rodríguez-Vázquez · J. Xamán · Y. Chávez
Centro Nacional de Investigación Y Desarrollo Tecnológico (Cenidet-TecNM-SEP), Prol. Av. Palmira S/N. Col. Palmira, Cuernavaca, Morelos, CP 62490, Mexico
e-mail: martinrovaz_iem@cenidet.edu.mx

J. Xamán
e-mail: jxaman@cenidet.edu.mx

Y. Chávez
e-mail: ycchena@cenidet.edu.mx

I. Hernández-Pérez (✉)
División Académica de Ingeniería Y Arquitectura (DAIA-UJAT), Universidad Juárez Autónoma de Tabasco, Carretera Cunduacán-Jalpa de Méndez Km. 1, Cunduacán, Tabasco, CP 86690, Mexico
e-mail: ivan.hernandezp@ujat.mx

F. Noh-Pat
Facultad de Ingeniería (UAC), Col. Buenavista, Universidad Autónoma de Campeche, Av. Agustín Melgar S/N, San Francisco de, Campeche, Campeche, CP 24039, Mexico
e-mail: felipe.noh@hotmail.com

Z. Driss et al. (eds.), *CFD Techniques and Thermo-Mechanics Applications*,
https://doi.org/10.1007/978-3-319-70945-1_3

Trombe wall, solar chimney, the wind tower, and earth–air heat exchanger (EAHE). The objective of an EAHE is to pre-cool or to pre-heat an air flow which is supplied to the interior of buildings, by taking advantage of the thermal inertia of the soil. In other words, in warm seasons the soil at certain depth has a lower temperature than ambient and higher temperature in cold season. The cooling and heating potential of the EAHE helps to improve the thermal comfort when used in not conditioned building or to reduce energy consumption when used in conditioned buildings.

The authors like (Mihalakakou et al. 1997) have used models to predict the thermal behavior of the soil through an energy balance. These models are still used to study EAHE. The EAHE cooling and heating potential was introduced in various studies such as Darkwa et al. (2010) and Ramírez-Dávila et al. (2014). Darkwa et al. (2010) evaluated the efficiency of an EAHE at the University of Nottinghan. The experimental results showed that the system provided a 62 and 86% of the cooling and the heating load of the building, respectively. Ramírez-Dávila et al. (2014) conducted a two-dimensional numerical study of the thermal performance of an EAHE in three cities of Mexico. Their results showed that in Juárez City and Mexico City, the EAHE performs better in summer than in winter, reducing air temperature up to 6.6 and 3.2 °C, respectively. While in Mérida, the EAHE presented a better performance in winter by increasing the temperature up to 3.8 °C.

The EAHE cooling and heating capacity varies depending on various factors including design, climatic conditions, type of soil, etc. Gauthier et al. (1997) studied the thermal behavior of an EAHE with the aim of reducing the energy consumption of greenhouses. The authors developed a parametric analysis of the key components in the EAHE design and operation. The results showed that as the distance between tubes and length increases, the amount of stored or recovered energy from the soil increases. On the contrary, the stored or recovered energy decreases as air speed increases. Finally, the results showed that the deeper the pipeline is buried, the greater the stored energy during the day is. Bansal et al. (2013) and Mathur et al. (2015) studied the effect of the thermal conductivity of the soil in the thermal behavior of EAHE. The authors showed that higher conductivity values give greater air temperature difference between the inlet and outlet the EAHE. Conversely, as the speed of the airflow increases, the air temperature difference decreases because the interaction time between the air and the soil decreases. Bisoniya et al. (2014) investigated the effect of different design parameters on the thermal performance of an EAHE for the city of Bhopal, India. They concluded that increasing the tube length causes pressure drops so more power is required by the fan to move the air. On the other hand, the EAHE outlet temperature tends to decrease as the diameter or the fluid velocity increases. In contrast, the air temperature at the outlet slightly increases as the depth increases. However, the excavation cost increases to a greater degree. Serageldin et al. (2016) conducted a parametric study of the thermal performance of an EAHE with different design parameters for the winter season in Egypt. The results of the study showed that the outlet temperature tends to decrease as the diameter or the fluid velocity increases. In contrast, the temperature tends to increase as the tube length increases. On the other hand, the distance between the pipes and tube material showed no significant effect on the thermal performance of the EAHE.

Works such as Brum et al. (2013), Dasare and Saha (2015) and Yang and Zhang (2015) have studied the effect of the depth to which the pipe is buried on the thermal behavior of EAHE. Brum et al. (2013) found that for the city of Viamao, Brazil, the variation of soil temperature decreases as the depth of burial is increased. However, from 3 m depth, no significant variation is noticed any longer. Dasare and Saha (2015) found that the optimal design of an EAHE for the city of Pune, India is a pipe with helical configuration (diameter spiral 0.2 m with spacing of 0.3 m) buried at a depth of 1.5 m, with an 0.25 m/s flow velocity double tube buried with a spacing of 1.2 m. Yang and Zhang (2015) observed that increasing the depth to which the pipes are buried (4 m maximum) improves the performance for EAHE in Chongquin, China. In their studies, the authors found an optimal burial depth at which the soil contribution in the thermal behavior of EAHE is negligible. This burial depth has proven to be different for each region, causing many authors to assume this depth relying on the values reported in the literature.

Currently, researchers are seeking to increase the cooling and heating capacity of EAHE by the coupling of other passive systems as a solar heater (Kaushal et al. 2015) or wind tower (Benhammou et al. 2015) which unified performance tends to improve the temperature variation more than that obtained with each system alone. Xamán et al. (2014) studied the effect of adding an insulation layer at the outlet section of an EAHE to minimize the losses or gains of energy through the interaction of the air with the earth at the outlet section. Based on this short review, we conclude that many factors influence the EAHE behavior. However, the geographical and climatic differences that exist in different regions may cause the EAHE optimal factors to be different. Then, it is convenient to perform a parametric study of EAHE design to maximize its heating and cooling capacity. This work aims to perform a parametric study to find the depth optimal values. The study was carried out using weather conditions in Ciudad Juarez, Chihuahua, Mexico City and Mérida, Yucatán, for the warmest and coldest days of the year.

1 Physical Model

Figure 1 presents the EAHE physical model. The following assumptions are taken into account:

1. Two-dimensional heat transfer.
2. Heat is gained or lost by radiation and convection on the surface.
3. Evaporation of water is only considered on the surface of the ground.
4. The thermal influence of the pipe is disregarded due to its thickness; it is so small that it is negligible. We have made this assumption because the thickness of a pipe is very small compared with the dimensions of the whole system, therefore the thermal resistance of the pipe wall can be neglected according to previous studies of earth-to-air heat exchangers (Bisoniya 2015; Yang and Zhang 2015).

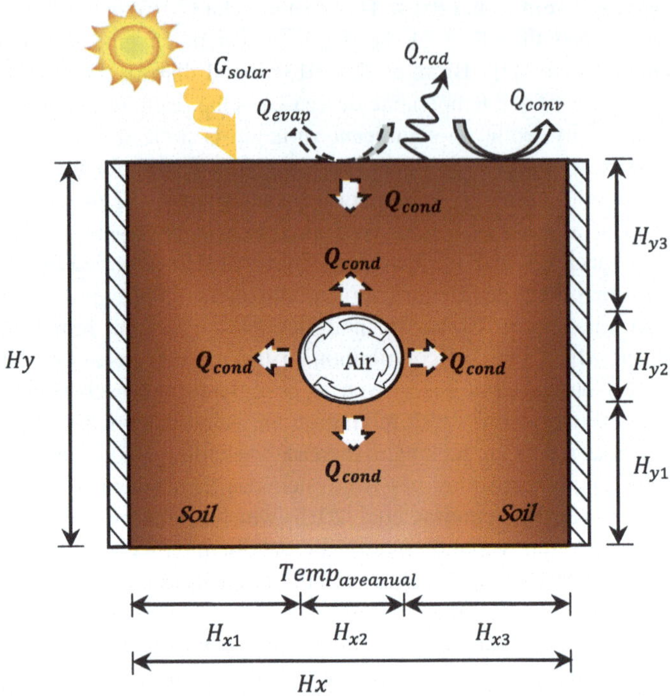

Fig. 1 The EAHE physical model

5. Condensation and evaporation inside of the pipe are not taken into account.
6. Inside the pipe the dominant heat transfer mechanism is convection and the air is in a laminar flow regime.
7. Interaction between the pipe and the soil is considered.
8. The thermophysical properties of the soil are constant.

Table 1 presents the dimensions considered to simulate the EAHE behavior.

Table 1 The EAHE dimensions

Section	Dimensions (m)
Depth from the surface to the pipe	Hy_3 = 0.425–4.425
Depth from the pipe to the bottom	Hy_1 = 4.425–0.425
Distance between the insulated wall and the pipe	Hx_1 = Hx_3 = 0.5
Pipe diameter	Hy_2 = Hx_2 = 0.15

2 Mathematical Model

The soil is considered a solid medium where heat is transferred by conduction. On the other hand, there is a convective heat transfer in laminar flow regime through the pipe, and a heat exchange between the walls of the pipe and the soil. These phenomena are modeled in the Cartesian coordinate system and described by continuity, momentum, and energy Eqs. (1–4).

$$\frac{\partial(\rho u)}{\partial x}+\frac{\partial(\rho v)}{\partial y}=0 \tag{1}$$

$$\frac{\partial(\rho u.u)}{\partial x}+\frac{\partial(\rho v.v)}{\partial y}=\frac{\partial}{\partial x}\left(\mu\frac{\partial u}{\partial x}\right)+\frac{\partial}{\partial y}\left(\mu\frac{\partial u}{\partial y}\right)-\frac{\partial P}{\partial x} \tag{2}$$

$$\frac{\partial(\rho u.u)}{\partial x}+\frac{\partial(\rho v.v)}{\partial y}=\frac{\partial}{\partial x}\left(\mu\frac{\partial u}{\partial x}\right)+\frac{\partial}{\partial y}\left(\mu\frac{\partial u}{\partial y}\right)-\frac{\partial P}{\partial y}-\rho g\beta(T-T_{\infty}) \tag{3}$$

$$\frac{\partial(\rho uT)}{\partial x}+\frac{\partial(\rho vT)}{\partial y}=\frac{\partial}{\partial x}\left(\Gamma\frac{\partial T}{\partial x}\right)+\frac{\partial}{\partial y}\left(\Gamma\frac{\partial T}{\partial y}\right) \tag{4}$$

The EAHE modeling consists in treating the geometry as a cavity with a fluid inside. In this cavity, the blocking technique assigns the value 0 to the velocities of the soil since only heat conduction is present in the soil. The governing equations for heat transfer and fluid flow were discretized with the finite volume method and the semi-implicit method. In order to couple the pressure, the SIMPLE algorithm was used together with a low-order scheme (hybrid). The boundary conditions for the above equations are as follows:

(a) East and west boundaries.

These boundaries are considered as adiabatic

$$\frac{\partial T}{\partial x}=0 \quad \text{at} \quad x=0 \quad \text{for} \quad 0\leq y\leq Hy \tag{5}$$

$$\frac{\partial T}{\partial x}=0 \quad \text{at} \quad x=Hx \quad \text{for} \quad 0\leq y\leq Hy \tag{6}$$

(b) South boundary.

The soil temperature is considered to be equal to the average annual ambient temperature.

(c) North boundary.

To represent the effect of conduction, convection, and radiation on the north boundary, the following energy balance developed by Mihalakakou et al. (1997) is implemented:

$$-q_{\text{cond}} = -q_{\text{conv}}(\text{CE}) + q_{\text{rad}}(\text{LR}) - \alpha G(\text{SR}) + q_{\text{evap}}(\text{LE}), \quad (7)$$

where

(i) CE is the convective energy exchange between the air and the surface of the ground:

$$q_{\text{conv}} = h_{\text{sur}}(T_{\text{amb}} - T_{\text{Sur}}), \quad (8)$$

where T_{amb} is the ambient temperature, T_{Sur} is the ground surface temperature. h_{sur} is the convective heat transfer coefficient at the ground surface, which can be calculated as

$$h_{\text{Sur}} = 5.678\left[0.775 + 0.35\left(\frac{\text{vel}_{\text{wind}}}{0.304}\right)\right] \quad \text{for} \quad \text{vel}_{\text{wind}} < 4.88 \quad (9)$$

$$h_{\text{Sur}} = 5.678\left[0.775 + 0.35\left(\frac{\text{vel}_{\text{wind}}}{0.304}\right)^{0.78}\right] \quad \text{for} \quad \text{vel}_{\text{wind}} \geq 4.88 \quad (10)$$

(ii) SR is a shortwave solar radiation absorbed by the surface of the ground.

where

$$\text{SR} = \alpha G \quad (11)$$

α is the ground absorptance and G is the incident solar radiation on the ground surface

(iii) LR is long wavelength radiation, which can be calculated as

$$\text{LR} = \varepsilon \Delta R \quad (12)$$

ε is the emittance of the ground surface and ΔR is a term that depends on the relative humidity of the ground and the outdoor air.

(iv) LE is the latent heat flux from the ground surface due to evaporation:

$$\mathrm{LE} = 0.0168 f h_{\mathrm{Sur}}[(aT_{\mathrm{Sur}} + b) - \mathrm{RH}(aT_{\mathrm{amb}} + b)] \tag{13}$$

where RH is the relative humidity on the surface of the ground and f is a fraction which depends on the type of ground cover. In this study, we used a value of 0.7 for f corresponding to the value of bare soil.

(d) Air at the inlet.

The air entering the pipe has the same temperature than the ambient and enters with at a constant speed, which is a function of the Reynolds number.

$$v = f(\mathrm{Re}), u = 0 \tag{14}$$

(e) Air at the outlet

$$\frac{\partial T}{\partial y} = 0, \frac{\partial u}{\partial y} = 0 \quad \text{and} \quad \frac{\partial v}{\partial y} = 0. \tag{15}$$

3 Methodology

The objective of this study is to find an optimal depth of burial for the pipe and also to analyze the behavior of the EAHE during a whole day. We studied nine depths of the tube (from 0.5 to 4.5 m, with increments 0.5 m).

To verify the computational code, a comparison of our results with the results reported by (House et al. 1990) (Fig. 2) was performed. The results were obtained for a Rayleigh number of 10^5 with conductivity ratios (CX) of 0.2 and 5. The geometry is a differentially heated cavity with adiabatic horizontal walls and west and east walls at temperatures T_H and T_C, respectively.

Table 2 presents the verification results where one can observe that all differences between the Nusselt numbers reported by House et al. (1990) and the Nusselt numbers obtained with our code are smaller than 1%.

After code verification, we carried out a study of grid independence. The medium was divided into 3 sections: one for the tube and two corresponding to the lower and upper regions of the tube (Fig. 2). The grid independence was performed using the climatic data of the warmest day of Juárez City for three burial depths (0.5, 2.5, and 4.5 m). The physical model was divided into three sections, and only the nodes number (Nx2 and Ny2) of the grid section corresponding to the pipe was varied. The other two sections have a fixed number of nodes. The grid independence study started with a grid of 11 × 11 for the pipe section and we increased the number of nodes until we had a grid with 71 × 71. The results show that for the air

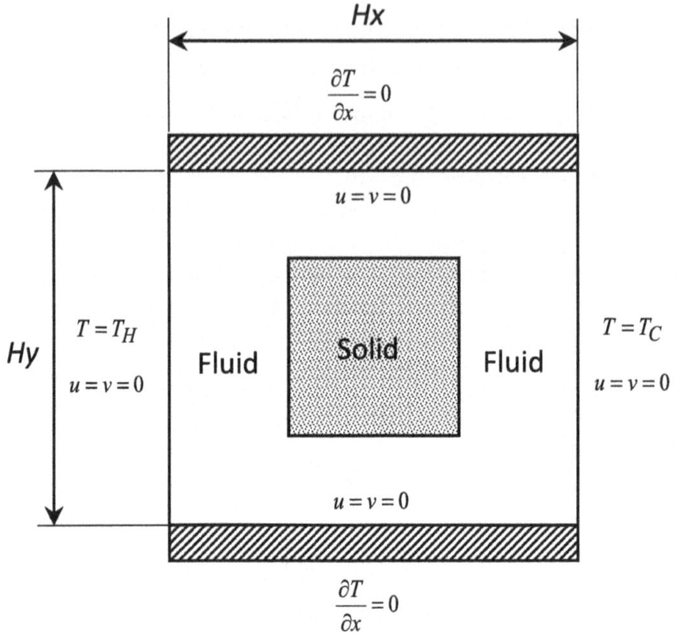

Fig. 2 Differentially heated cavity with an embedded solid

Table 2 Average Nusselt values at the hot wall for conductivity reasons of 0.2 and 5

CX	House et al. (1990)	Current work	Absolute difference (%)
0.2	4.62	4.65	0.65
5	4.32	4.35	0.69

Table 3 Average air temperature and average enthalpy difference for different grid sizes in the section of the tube

Grid size	Average temperature			Average enthalpy difference		
	0.5 m	2.5 m	4.5 m	0.5 m	2.5 m	4.5 m
11 × 11	28.9875	24.4354	19.8912	−7241.3008	−11811.6748	−16374.0396
21 × 21	28.9811	24.4133	19.8830	−7247.7678	−11833.8141	−16382.2983
31 × 31	28.9787	24.4041	19.8794	−7250.1539	−11843.0706	−16385.8921
41 × 41	28.9788	24.4045	19.8781	−7250.0429	−11842.7267	−16387.1955
51 × 51	28.9787	24.4047	19.8796	−7250.2243	−11842.4331	−16385.7194
61 × 61	28.9788	24.4051	19.8795	−7250.1180	−11842.1179	−16385.7821
71 × 71	28.9789	24.4054	19.8795	−7250.0116	−11841.8026	−16385.8193

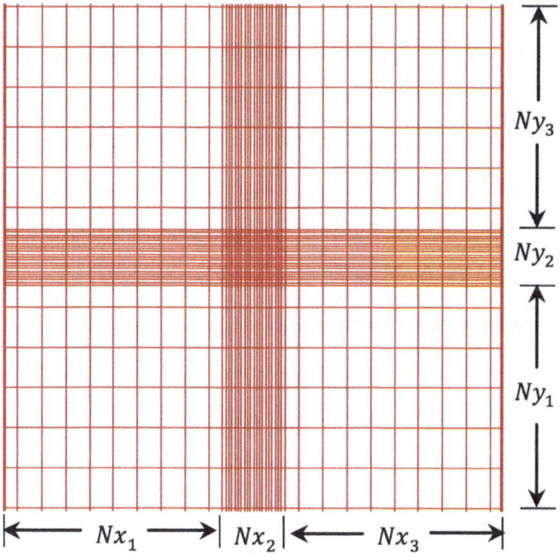

Fig. 3 Non-uniform grid

temperatures and enthalpies the difference between 51 × 51-node grid the 61 × 61-node grid there was not significant; here was a maximum difference of 0.01%. Table 3 indicates the results for the grid independence (Fig. 3).

4 Results and Discussion

The results of the parametric study to find the burial optimal depth are presented below. Tables 4 and 5 present the input values of radiation, ambient temperature, relative humidity and wind speed for the three cities considered in this study. Climatic data correspond to the coldest and warmest hour of winter and summer days, respectively.

Figure 4 shows that the air temperature inside the tube decreases when increasing depth. Among the cities, Mérida witnessed the greatest decrease; in this city, the temperature drops 1.6 °C for every 0.5 m of depth increment, while for Juárez City and Mexico City the temperatures decrease reached 1.1 and 1.3 °C, respectively.

Table 4 Climatic conditions of the three studied cities for the hottest hour of the year

City	Day/month	Hour (pm)	G (W/m^2)	T (°C)	RH (%)	Wind speed (m/s)
Juárez City	27/07	14:00	872	36.2	30	7.4
México City	20/05	13:00	930	28.9	23	2.3
Mérida	20/05	12:00	1135	39.8	37	2.4

Table 5 Climatic conditions of the three studied cities for the coldest hour of the year

City	Day/month	Hour (am)	G (W/m^2)	T (°C)	RH (%)	Wind speed (m/s)
Juárez City	13/01	4:00	0	−3.7	69	3.1
Mexico City	13/01	7:00	0	1.8	61	2.7
Mérida	12/01	7:00	0	10.1	100	1.6

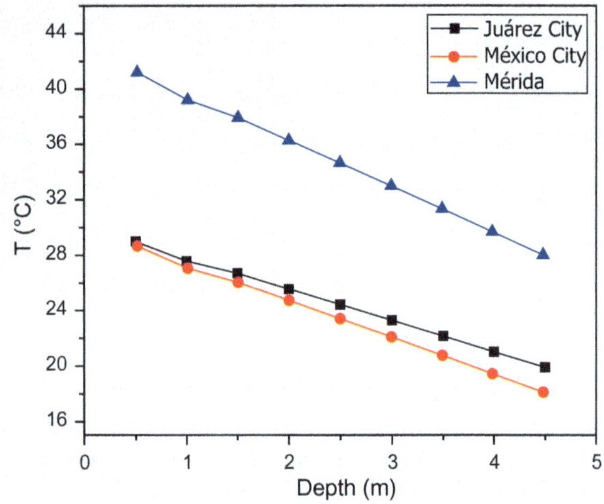

Fig. 4 Average temperature inside the pipe for different burial depths on a warm day

Thus, a temperature difference between the smallest depth (0.5 m) and the biggest (4.5 m) reaches 13.3, 10.6, and 9.1 °C for Mérida, Mexico City, and Juárez City, respectively. Figure 5 presents the air temperature profiles for different depths for Mérida.

Figure 6 presents the thermal performance of EAHE in the heating function. The temperature increases in greater proportion in Juárez City, an average of 2.4 °C per each 0.5 m of depth increment. In the case of Mexico City and Mérida, the temperature increases 1.8 °C per 0.5 m of depth increment in both cities. Thus, between the minimum and maximum depths a temperature difference of 19.2, 14.3 and 14.5 °C was noted for Juárez City, Mexico City, and Mérida, respectively.

For the heating operation, the highest temperature increase occurred in Juárez City. However, the highest air temperature inside the tube occurs in Mérida. This can be explained by the fact that the temperature of the soil and the ambient temperature are higher in Mérida compared to Juárez City and Mexico City. Figure 7 presents the air temperature profiles for different depths for Juárez City.

Figure 8 shows the amount of energy lost by the air in EAHE during the warm season. The energy loss (change of enthalpy) from the air is proportional to its temperature drop. Similar to the analysis of temperature, the greatest energy loss occurred in Mérida, with an average of −1671 kJ/kg for every 0.5 m of depth

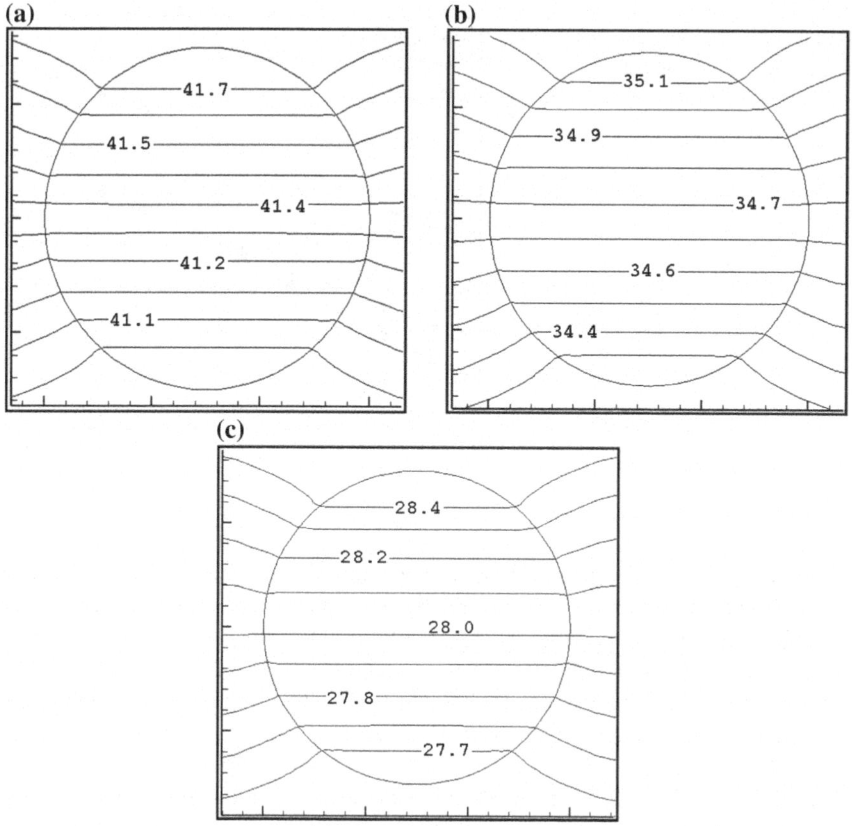

Fig. 5 Average temperature profile inside the pipe for Mérida on a warm day: **a** 0.5 m, **b** 2.5 m, and **c** 4.5 m

increment. In Juárez City and Mexico City, the loss of energy averaged −1142 and −1336 kJ/kg, respectively.

Figure 9 presents the heat gain of air at different burial depths for the operation of EAHE in a cold season. As in the case of temperature, the highest value of gained energy occurred in Juárez City, with an average of 2408 kJ/kg. The average change enthalpy for Mexico City and Mérida is 1792 and 1820 kJ/kg, respectively.

Figure 10a, b show the average values of air velocity and average Reynolds number, respectively. These figures show that the air velocity and Reynolds number for the cold season are very similar regardless of the depth to which the tube is buried and it is the city of Juarez that displayed the highest air velocities.

Using the optimal depth, which was 4.5 m for all the cities, we have analyzed the behavior of the cooling and heating potential of the EAHE for a whole day for Juárez City using four different Reynolds numbers for the 2-D geometry presented in (Xamán et al. 2014). We have selected this city because it is the one with the most extreme climate. Table 6 gives the climatic data used for the simulation of the

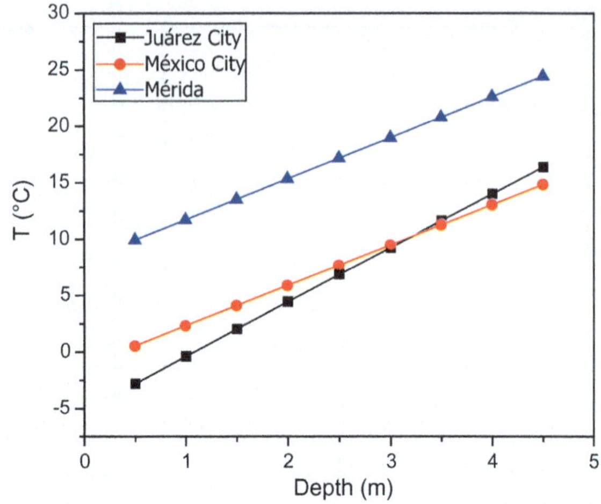

Fig. 6 Average temperature inside the pipe for different burial depths on a cold day

coldest and warmest day of the year. Figure 11 shows that the EAHE has a good cooling potential. The behavior of the air temperature at the outlet during the whole day is similar to that of the outdoor air, but the temperature at the outlet is lower than the outdoor air temperature by at least 5.9 °C. Therefore, the minimum cooling effect of the EAHE during the day is 5.9 °C and it occurs at 9:00 h for all Reynolds.

Figure 12 shows the cooling effect, which is the difference of temperature between the outlet and the inlet. For the remaining hours, either before or after 9:00 h, the EAHE has a better cooling potential. During the first 4 h of the day, from 1:00 to 4:00 h, the air temperature at the outlet remains almost constant for each Reynolds number. The biggest cooling effect during these hours is equal to 8.8 °C and it is provided by a Re = 100. Then at 5:00 h the air temperature at the outlet decreases because the wind velocity has a sharp decline from 2.3 to 0.8 m/s (Fig. 4b). The biggest cooling effect during this hour is 9.6 °C for Re = 100. After 5:00 h the cooling potential of the EAHE decreases each hour until it reaches its minimum value. After the 9:00 h, the cooling potential improves considerably; it increases each hour until it reaches its maximum that occurs at 21:00. For Re = 100 the EAHE is able to cool the air by 17.4 °C below the temperature that it has at the inlet. For Re = 500, 1000, 1500 the maximum cooling potential of the EAHE is 16.4, 15.5 and 14.8 °C, respectively. During the hours with solar irradiation (from 8:00 to 19:00 h), the air temperature at the outlet of the EAHE is similar for all Re, the maximum difference is just 1 °C. Therefore, for cooling purposes it is recommended to use the EAHE with a Reynolds equal to 100.

For heating purposes, the EAHE worked correctly for all hours and all Reynolds during the cold day (except at 18:00 and 19:00, Re = 100) in Juárez City. Figure 13 indicates that the EAHE has a better heating potential with a high Reynolds number. Just like what occurred on a warm day, the behavior of the air temperature graph at the outlet during the whole day is similar to that of the outdoor air, but the outlet temperature is greater than the outlet air temperature. Figure 13 shows the

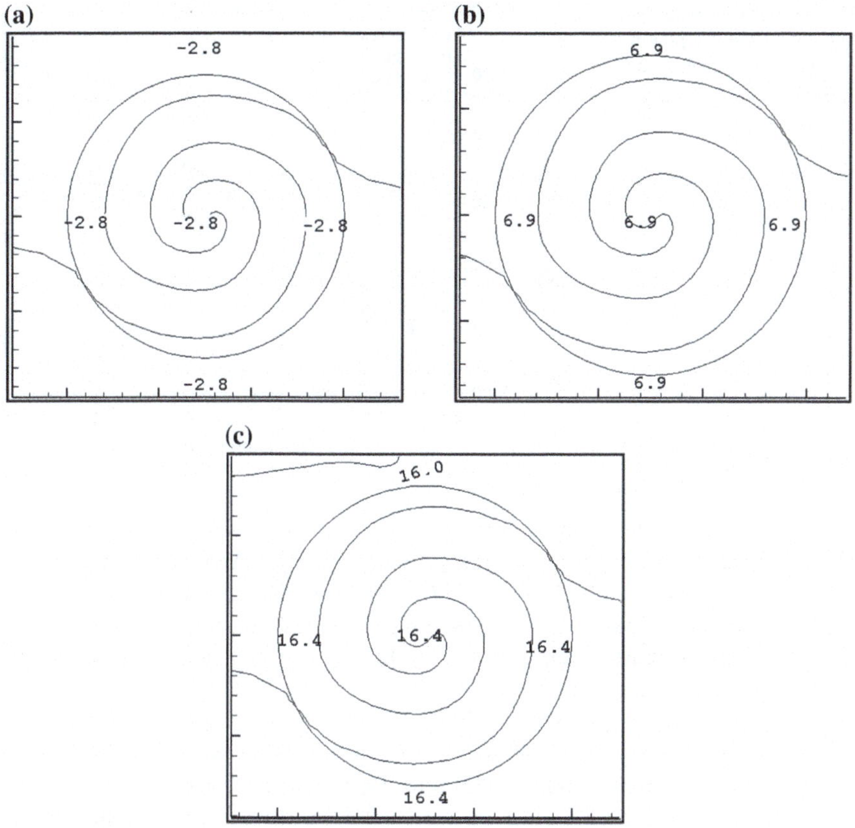

Fig. 7 Average temperature profile inside the pipe for Júarez City on a cold day: **a** 0.5 m, **b** 2.5 m, and **c** 4.5 m

Fig. 8 Average change of enthalpy inside the pipe for different burial depths on a warm day

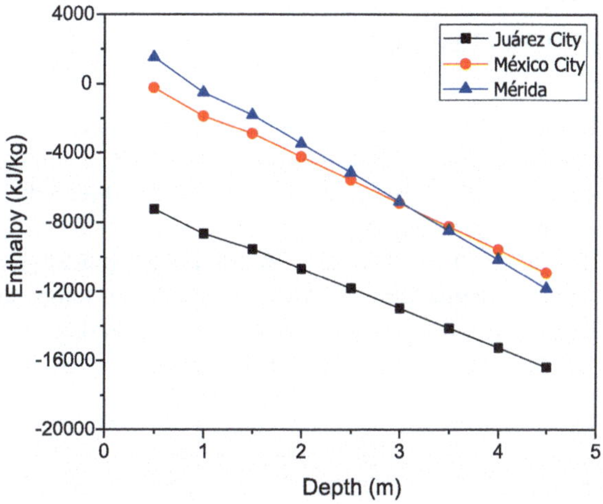

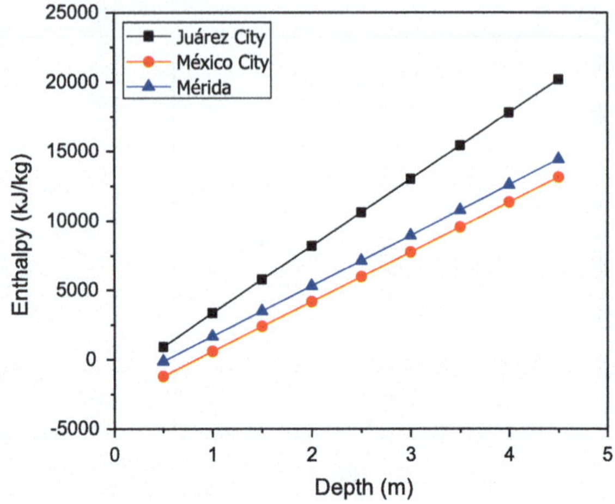

Fig. 9 Average change of enthalpy inside the pipe for different burial depths on a cold day

heating potential, which is the difference of temperature between the outlet and the inlet. The maximum heating potential provided by the EAHE during the day is 6.3 °C and it occurs at 4:00 h. From 5:00 to 10:00 h, the temperature of the air at the outlet remains almost constant for each Reynolds number; for example, for Re = 1500 the heating potential is 6.0 °C for these 5 h. After the 10:00 h, the heating potential decreases to its minimum value (1.4 °C) at 18:00 h, for Re = 1500. During the period from 10:00 to 18:00 h, the graphs of the outlet air temperature are close to each other than during the rest of the day. Then, the heating potential improves from 19:00 to 24:00 h for Re = 500, 1000 and 1500. Again, the Re = 1500 provides the best heating potential than other Reynolds number. As seen, the heating potential of the EAHE is lower than the cooling potential, but the former is still important because this city has very low temperatures in winter.

5 Conclusions

The aim of this work was to simulate numerically the variation of air temperature inside an EAHE at different burial depths. The simulations were carried out for the warmest and coldest hours of the year for three cities in Mexico.

On a warm day, the depth at which the tube EAHE is buried has a very significant effect in all cities; in Mérida, the temperature decreases 1.6 °C each 0.5 of depth, in Juárez City and Mexico City the air temperature decreases 1.1 and 1.3 °C each 0.5 m of depth, respectively. On a cold day, the burial depth has a noticeable effect on Juárez City by a (2.4 °C) air temperature increment for every 0.5 m burial

Fig. 10 Average values inside the pipe for different burial depths on a cold day: **a** Average air speed and **b** Average Reynolds number

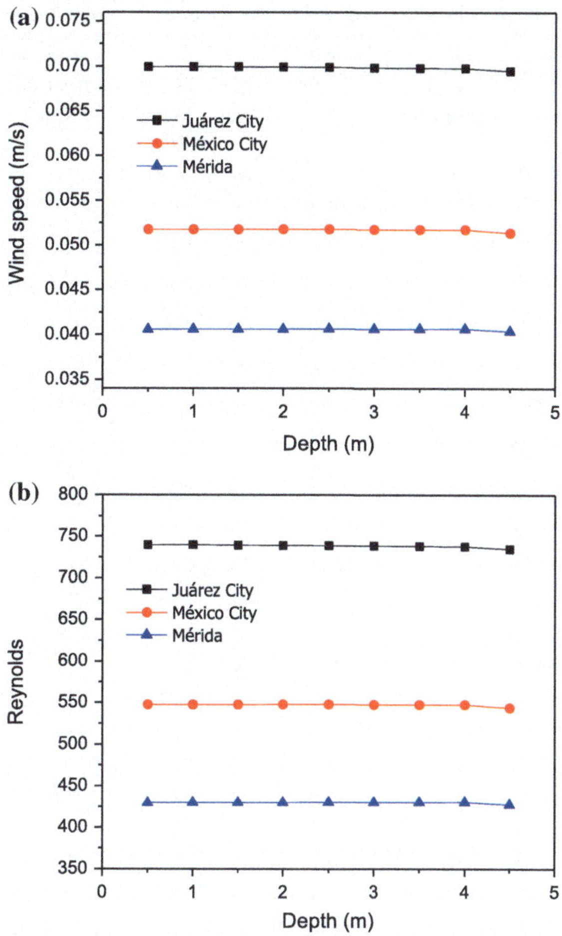

depth compared to Mexico City and Mérida (1.8 °C in both cities). Despite this large temperature difference between depths, the highest temperature was presented under the climatic conditions of Mérida. However, the air supplied to the three cities EAHE has a relatively high temperature, compared to room temperature. Thus, the EAHE works well for the three cities in the heating function. Considering the temperature variation for each 0.5 m in the three cities, it is concluded that increasing the depth more than 5 m is unnecessary, since after this depth the air temperature is similar to soil temperature.

Using the optimal depth, we analyzed the behavior of the cooling and heating potential of the EAHE for a whole day for Juárez City. In this city, the EAHE

Table 6 Climatic conditions for the warmest and coldest days in Júarez City

Time	Coldest day (January 13)				Warmest day (July 27)			
	Irradiance (W/m^2)	T (°C)	RH (%)	Wind speed (m/s)	Irradiance (W/m^2)	T (°C)	RH (%)	Wind speed (m/s)
1:00	0	−3.7	69	0.9	0	25.1	52	1.9
2:00	0	−3.7	69	1.0	0	24.9	51	2.9
3:00	0	−3.7	68	1.8	0	24.8	53	4.8
4:00	0	−3.7	69	3.1	0	24.7	52	5.3
5:00	0	−3.6	71	1.8	0	24.6	53	4.0
6:00	0	−3.6	72	2.5	0	24.6	55	3.5
7:00	0	−3.6	72	3.1	9	24.8	58	4.0
8:00	0	−3.5	76	2.7	158	26.4	52	4.0
9:00	17	−2.9	73	2.7	351	28.4	45	4.5
10:00	184	0.1	57	3.5	522	30.4	40	4.8
11:00	332	2.9	47	4.2	676	32.2	36	5.5
12:00	447	5.4	39	3.5	780	33.9	34	6.0
13:00	511	7.3	35	3.8	810	35.1	31	7.0
14:00	522	8.7	31	3.8	872	36.2	30	7.4
15:00	481	9.5	27	3.3	829	37.0	28	8.1
16:00	390	9.7	26	3.0	723	37.3	27	8.4
17:00	253	9.0	27	3.1	597	37.4	28	6.3
18:00	97	7.5	30	3.5	393	36.8	29	5.8
19:00	1	5.8	35	4.8	223	35.8	31	6.8
20:00	0	5.2	43	3.8	60	34.5	34	7.4
21:00	0	4.7	43	2.5	0	33.2	39	6.8
22:00	0	4.1	49	2.1	0	32.7	40	5.5
23:00	0	3.5	49	3.3	0	32.2	42	5.5
24:00	0	3.0	49	2.3	0	31.8	44	5.8

worked correctly for cooling purposes during the whole day. The EAHE has a good cooling potential in this city because it is able to cool the air up to 17.4 °C below the temperature at the inlet. Such cooling effect can help to improve the thermal comfort inside a building located in this city because the air temperature at outdoors can reach 42 °C during the day. On the other hand, the maximum heating effect provided by the EAHE on the coldest day is 6.3 °C. The heating potential of the EAHE is not as good as the cooling potential but it is still important because this city has low temperatures up to −3.7 °C on the coldest day.

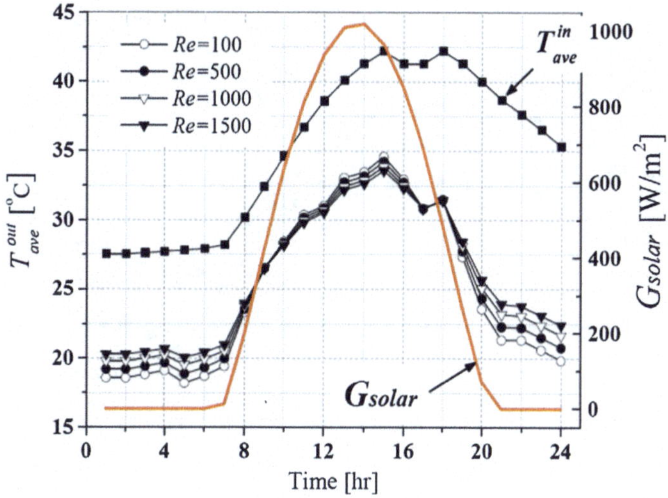

Fig. 11 EAHE Behavior along the day in Júarez City for cooling

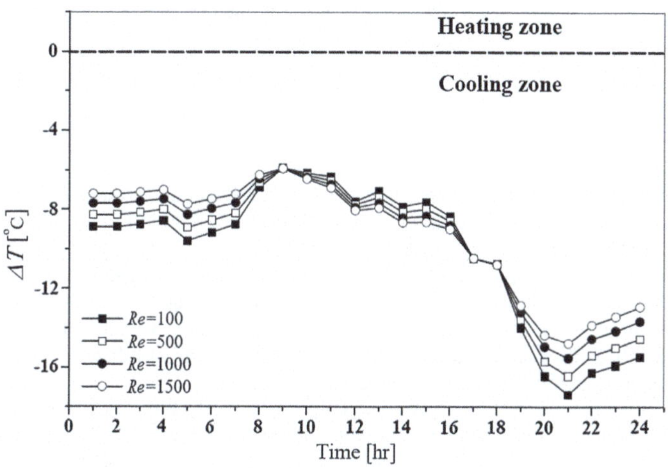

Fig. 12 Behavior of the EAHE during the day in Júarez city for cooling

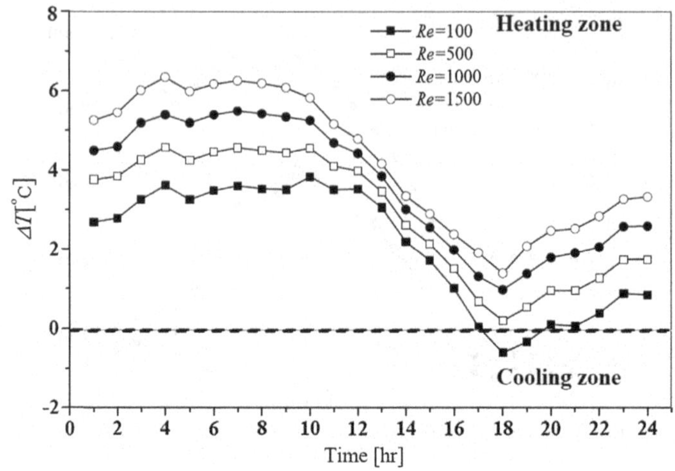

Fig. 13 Behavior of the EAHE during the day in Júarez city for heating

Acknowledgements M. Rodríguez-Vázquez acknowledges the support provided by the Consejo Nacional de Ciencia Y Tecnología (*CONACYT*) given through its doctorate scholarship program.

References

Bansal V, Misra R, Agarwal GD, Mathur J (2013) Transient effect of soil thermal conductivity and operation on performance of earth-air tunnel heat exchanger. Appl Energy 103:1–11. https://doi.org/10.1016/j.apenergy.2012.10.014

Benhammou M, Draoui B, Zerrouki M, Marif Y (2015) Performance analysis of an earth-to-air heat exchanger assisted by a wind tower for passive cooling of building in arid and hot climate. Energy Convers Manag 91:1–11. https://doi.org/10.1016/j.enconman.2014.11.042

Bisoniya TS (2015) Design of a earth-air heat exchanger system. Geothermal Energy 3:18–27. https://doi.org/10.1186/s40517-015-0036-2

Bisoniya TS, Kumar A, Baredar P (2014) Heating potencial evaluation of earth-air heat exchanger system for Winter season. J Building Phys 39:1–19. https://doi.org/10.1177/1744259114542403

Brum RS, Vaz J, Rocha LAO, dos Santos ER, Isoldi LA (2013) A new computational modeling to predict the behavior of earth-air heat exchangers to predict the behavior of earth-air heat exchangers. Energy Build 64:395–402. https://doi.org/10.1016/j.enbuild.2013.05.032

Darkwa J, Kokogiannakis G, Magadzire CL, Yuan K (2010) Theoretical and practical evaluation of an earth-tube (E-tube) ventilation system. Energy Build 43:728–736. https://doi.org/10.1016/j.enbuild.2010.11.018

Dasare RR, Saha SK (2015) Numerical study of horizontal ground heat exchanger for high energy demand applications. Appl Therm Eng 85:252–263. https://doi.org/10.1016/j.applthermaleng.2015.04.014

Gauthier C, Lacroix M, Bernier H (1997) Numerical simulation of soil heat exchanger storage system for greenhouses. Sol Energy 60:333–346. https://doi.org/10.1016/S0038-092X(97)00022-4

House JM, Beckermann C, Smith TF (1990) Effect of a centered conducting body on natural convection heat transfer in an enclosure. Numer Heat Transfer 18:225–231. https://doi.org/10.1080/10407789008944791

Kaushal M, Dhiman P, Singh S, Patel H (2015) Finite volume and response surface methodology based performance prediction and optimization of a hibryd earth to air tunnel heat exchanger. Energy Build 104:25–35. https://doi.org/10.1016/j.enbuild.2015.07.014

Mathur A, Srivastava A, Mathur J, Agrawal GD (2015) CFD analisys of EATHE system under transient conditions for intermittent operation. Energy Build 87:37–44. https://doi.org/10.1016/j.enbuild.2014.11.022

Mihalakakou G, Santamouris M, Lewis JO, Asimakopoulos DN (1997) On the application of the energy balance equation to predict ground temperatures profiles. Sol Energy 60:181–190. https://doi.org/10.1016/S0038-092X(97)00012-1

Ramírez-Dávila L, Xáman J, Arce J, Alvarez G, Hernandez-Pérez I (2014) Numerical study of earth to air heat exchanger for three different climates. Energy Build 76:238–248. https://doi.org/10.1016/j.enbuild.2014.02.073

Serageldin AA, Abdelrahman AK, Ookawara S (2016) Earth-air heat exchanger thermal performance in Egypcian conditions: experimental results, mathematical model, and computational fluid dynamics simulation. Energy Convers Manag 122:25–38. https://doi.org/10.1016/j.enconman.2016.05.053

Xamán J, Hernández-Peréz I, Arce J, Álvarez G, Ramírez-Davíla L, Noh-Pat F (2014) Numerical study of earth to air heat exchanger: the effect of thermal insulation. Energy Build 85:356–361. https://doi.org/10.1016/j.enbuild.2014.09.064

Yang D, Zhang J (2015) Analysis and experiments on the periodically fluctuating air temperature in a building with earth air tube ventilation. Build Environ 85:29–39. https://doi.org/10.1016/j.buildenv.2014.11.019

CFD Modeling of a Parabolic Trough Receiver of Different Cross Section Shapes

Anissa Ghomrassi, Hatem Mhiri and Philippe Bournot

1 Introduction

Solar energy is the most important renewable energy source that can significantly contribute to satisfy an increasing energy demand. Concentrated solar power (CSP) is considered as one of the proven energy conversion procedures. Among the existing CSP systems (PTC, solar tower, reflector Fresnel mirrors, and solar dish), the PTC technology is one of the most relevant and mature technologies for solar energy exploitation. It is also classified as proven and low cost large-scale solar power existing technology (Mao et al. 2014).

The PTC is essentially composed of a reflector and a receiver tube. The parabolic-shaped mirrors concentrate incident solar radiation on the lowest half absorber tube periphery. The other half of the receiver periphery is subjected to direct radiation transmitted through the glass envelope. The absorber converts the concentrated flux to heat and transfers it after that to the working fluid circulating through the receiver tube. This latter is generally covered with a glass envelope with the aim of reducing heat losses to the surrounding environment.

In order to better evaluate the optical performance of this solar energy conversion system, it is essential to evaluate concentrated solar heat flux and estimate its distribution on the receiver tube periphery. To this end, an analytical semi-finite formulation of the PTC concentrated heat flux density was developed by

A. Ghomrassi (✉) · H. Mhiri
Unit of Thermal and Thermodynamics in Industrial Processes,
National Engineering School of Monastir, Monastir, Tunisia
e-mail: ghomrassi.anissa@hotmail.com

H. Mhiri
e-mail: hatem.mhiri@enim.rnu.tn

P. Bournot
IUSTI, UMR CNRS 6595, Technopole de Chateau-Gombert, Paris, France
e-mail: philippebournot@yahoo.fr

Z. Driss et al. (eds.), *CFD Techniques and Thermo-Mechanics Applications*,
https://doi.org/10.1007/978-3-319-70945-1_4

Jeter (1986). Jeter was the first to study the PTC system and determine the concentrated solar heat flux distribution on the absorber tube periphery. The analytical method established by Jeter (1986, 1987) was able to evaluate the PTC system optical efficiency. The solar power distributions on the PTC focal plane for different incidence angles and for different opening radii are determined and compared as shown in both Jeter's publications (Jeter 1986, 1987). There are no bibliographic reviews of experimental work measuring solar flux distributions on the PTC focal line. That is why it is necessary to estimate it by numerical investigations. The Monte Carlo ray tracing method (MCRT) is frequently utilized to determine the concentrated solar heat flux distributions on the absorber. This technique had been largely used in earlier studies concerning tower plants and parabolic dish. Delatorre et al. (2014) proposed a significant database detailing the MCRT technique. Bibliographic references focusing on PTC systems are rare. In fact, among the recent literature reviews, a numerical coupled study combining the MCRT method and the FLUENT software was carried out by Wang et al. (2010a). A tridimensional numerical simulation has been developed to estimate concentrated solar heat flux profiles, temperature distribution profiles, and thermal stresses values for both eccentric and concentric receiver tubes. Wang et al. (2010b) have also studied the receiver tube material selection effect on tubes thermal stress.

Because the absorber is the PTC essential element, numerous literature studies have been focused on the PTC tube (Ouagued et al. 2013; Ramchandra et al. 2014). The manufacturing and maintenance cost of the receiver tubes is approximately 20% higher than of the parabolic trough concentrator whole cost. Consequently, this study spotlights the absorber tube geometry selection and optimization in order to ameliorate this solar conversion system performance. This paper examines the effect of a cross section shape modification. The comparison between the different shapes tested allows us to conclude the advantages and disadvantages of each shape.

2 Solar Heat Flux Calculation

The PTC geometric parameters are recapitulated in Table 1 (Wang et al. 2010a, b). The reflector material is aluminum while the absorber material is stainless steel coated with "black chromium" and covered by a glass envelope. PTC components optical properties are recapitulated in Table 2. The heat flux density calculation was performed by SOLTRACE for the 21st June at 12 h local time for 39.5 latitude and 116.2 longitude. It corresponds to a direct normal irradiance (DNI) of 1000 W/m^2. The sun shape is recommended as pillbox distribution with a half-width of 4.65 mrad (Sargent and Lundy 2003) (Fig. 1).

Table 1 Geometric parameters used in the simulation

PTC system	Value
Aperture width, w_a	1 m
Focal length, f	2 m
Collector length, L	2 m
Inner tube diameter, d_{ri}	0.06 m
Outer tube diameter, d_{ro}	0.07 m
Glass envelope inner diameter, d_{gi}	0.1 m
Glass envelope outer diameter, d_{go}	0.105 m
Opening radius	45°

Table 2 Optical proprieties of PTC components (Wang et al. 2010b)

	Absorptivity	Reflectivity	Transmissivity
Receiver tube	0.95	0	0
Glass envelope	0.01	0.01	0.965
Reflector	0	0.95	0.05

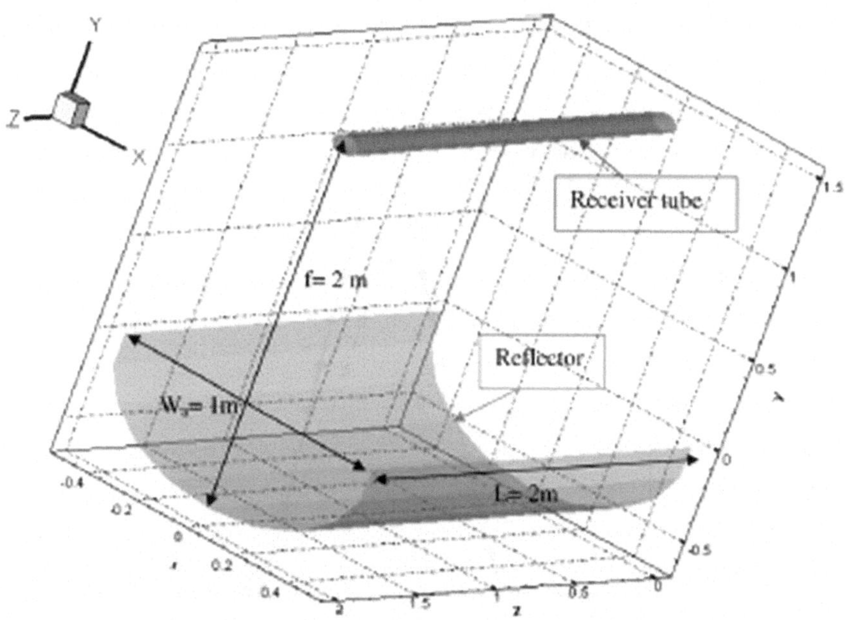

Fig. 1 Schematic of studied parabolic trough collector

2.1 Radius Number Sensitivity

The ray number traced through the system affects significantly the accuracy and the distribution of the solar heat flux on the absorber tube periphery. For this, a ray number sensitivity study was conducted. The ray numbers tested were 10^4, 10^5, 10^6, and 10^7. The results show that the ray numbers 10^5, 10^6, and 10^7 give similar results. On the other hand, the ray number of 10^4 gives a different solar heat flux distribution. That is why we opted for a 10^5 ray number for all simulations, which allows us to save time in simulations of calculation compared to the numbers 10^6 and 10^7. Figure 2 depicts the solar heat flux distributions on the absorber tube periphery for different tested ray numbers.

2.2 Validation

The validation of this modeling approach was achieved by comparing the numerical results calculated by SOLTRACE and Jeter (1986) data. The local concentration ratio distribution (LCR = q/DNI) at the PTC focal plane is compared to the LCR distribution given by Jeter (1986) as shown in Fig. 3. A good concordance is obtained between numerical and analytical data. The results are statistically significant with very high R-squared value (above 0.987) and very low *P*-value

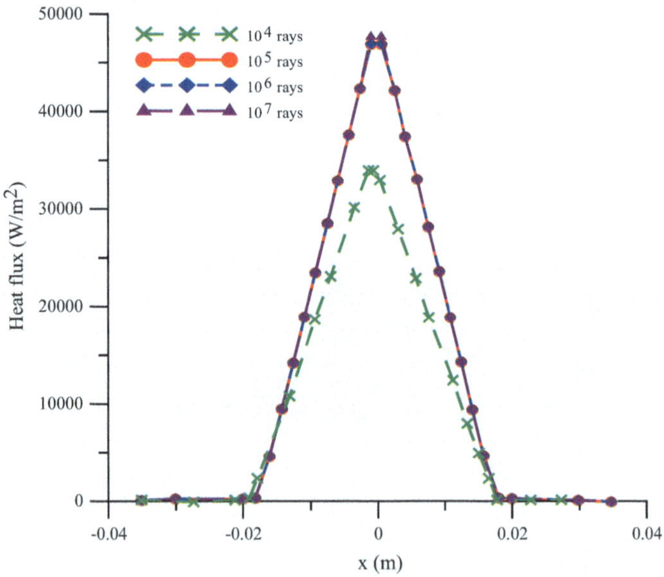

Fig. 2 Ray numbers effect on the solar flux distribution

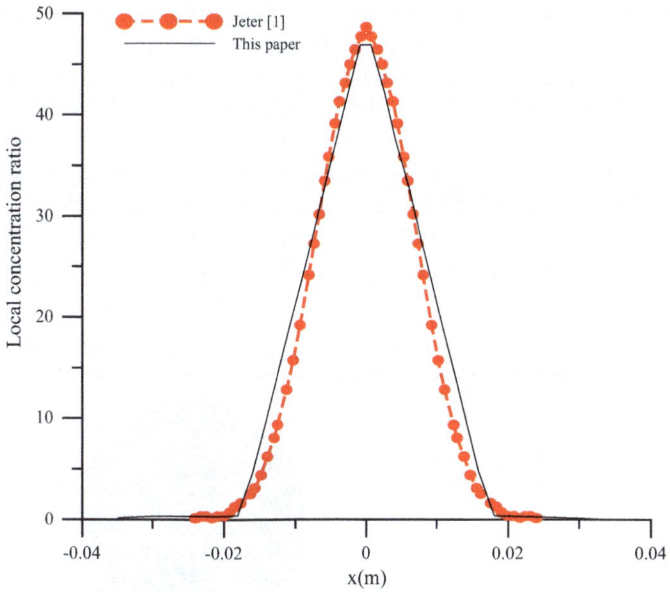

Fig. 3 Comparison of SOLTRACE modeling results and Jeter's (1986) data

(in a range of $0 < P\text{-value} < 10^{-7}$). These results confirm that the MCRT method employed in the present paper is appropriate for such problem.

3 CFD Numerical Simulation

Concentrated solar heat flux distributions estimated by the MCRT described above are imported into FLUENT software as wall thermal boundary conditions. The numerical simulations allow us to analyze the dynamic and thermal behavior of working fluid circulating in the absorber tube.

3.1 Geometry Configuration and Mesh

The geometry configuration of the PTC was carried out by Gambit software (GAMBIT 2005). The configuration is detailed in Fig. 1, and the corresponding parameters are presented in Table 1. Quadrilateral nonuniform meshes were chosen for rectangular and circular tube cross sections, and an unstructured triangular mesh was generated for the triangular cross section. For the three configurations tested, the mesh is finer in the transversal (angular), radial sections and also near the walls. This nonuniformity is explained by the large difference of the solar heat flux distributions illustrated in both angular and radial directions. A grid dependency study has been carried out since the mesh density affects considerably the numerical results accuracy. It has been ensured that numerical results are independent from cell numbers of 765,440; 810,000; and 864,000 control volumes for the cylindrical, rectangular, and triangular cross sections absorber tubes, respectively (Fig. 4). Given the need for a high resolution near the walls, an Enhanced wall treatment model was used with the nondimensional distance near the walls $y^+ \approx 1$ in all simulations (Dongqiang et al. 2013). After that, these meshed configurations were exported to the FLUENT software (Fluent 2005) for resolution.

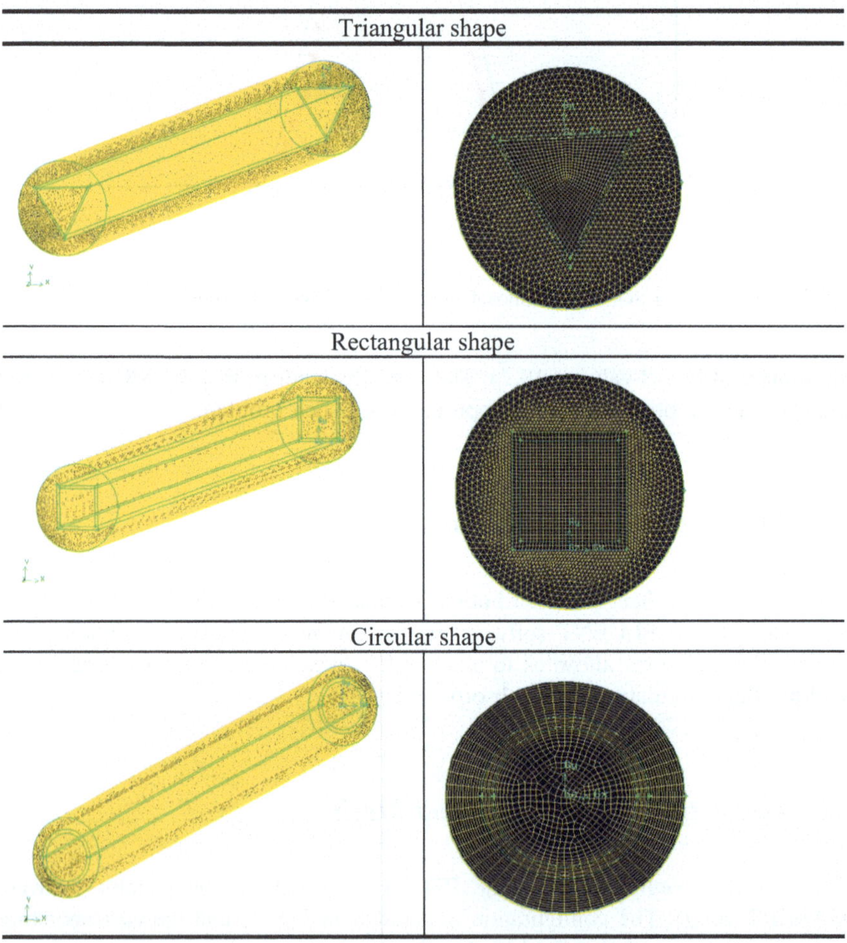

Fig. 4 Longitudinal and transversal mesh details for different configurations of receiver tubes tested

3.2 Assumptions and Boundary Conditions

The considered assumptions are the following:

(a) The flow is turbulent and in a steady state.
(b) For this type of flow, where solar heat flux gradient between the tube lower part (concentrated heat flux) and upper part (1000 W/m^2) is high, the convection is considered as mixed convection.
(c) The annular space is under vacuum, meaning that the radiative heat transfer is considered between the outer tube surface and the inner glass envelope surface.
(d) Thermal oil fluid (Syltherm 800) is the working fluid circulating in the absorber tube. The working fluid physical properties are temperature dependent. They are considered as a polynomial function of temperature (Cheng et al. 2012).

The boundary conditions applied to the domain frontiers are defined as follows:

(a) At the inlet of the tube, the mass flow inlet of the working fluid is fixed to 0.374 kg/s at a temperature of 300 K.
(b) At the outlet of the tube, a fully developed flow is considered:

$$\frac{\partial u}{\partial x} = \frac{\partial v}{\partial x} = \frac{\partial w}{\partial x} = \frac{\partial k}{\partial x} = \frac{\partial \varepsilon}{\partial x} = \frac{\partial p}{\partial x} = 0 \tag{1}$$

(c) On the outer glass envelope, the heat transfer with the surrounding external environment is achieved by mixed convection and radiation. This boundary condition introduces two temperatures: ambient temperature (T_a) and sky temperature (T_{sky}) connected by the following equation (Ghomrassi et al 2015):

$$T_{sky} = 0.0552\, T_a^{1.5} \tag{2}$$

The heat transfer coefficient around the glass envelope is calculated from the following correlation (Ghomrassi et al. 2015):

$$h_w = 4 v_w^{0.58} d_{go}^{-0.42} \tag{3}$$

where v_w is the wind velocity, and d_{go} is the outer diameter of the glass envelope.
(d) The absorber tube lower half periphery exposed to the reflector is subjected to a nonuniform solar heat flux distribution calculated by the SOLTRACE software. The solar heat flux distribution calculation was detailed in Sect. 2.
(e) The top half periphery is exposed to an incident solar heat flux of 1000 W/m^2.

3.3 Numerical Simulation

The turbulent flow modeling in steady state was ensured by solving differential equations expressing the continuity, momentum, and energy conservations. These equations closure was performed using the k-ε RNG turbulence model associated with the wall treatment. In fact, for regions close to solid walls where viscous effects predominate the turbulence effects, the near wall model approach "enhanced wall treatment" was used to handle near wall regions (Dongqiang et al. 2013; Usta 2010).

The first-order upwind discretization scheme was chosen for all equations except for the "momentum" and "energy" where the second-order upwind scheme was used (Cheng et al. 2012). The SIMPLE algorithm was used to allow the coupling between pressure and velocity (Patankar and Spalding 1972).

The thermal radiation heat transfer between the outer absorber tube wall and the inner glass envelope tube surface was also taken into account by choosing DO radiation model (discrete ordinates model). This radiation model is characterized by the angular discretization in space.

The convergence criterion determined for all the flow variables is defined as the maximum residue divided by the first maximum residue of five iterations cells. It is less than 10^{-5} for the momentum and continuity and less than 10^{-7} for the energy and DO radiation model.

3.4 CFD Simulation Results

In this paper, we were interested in the effect of the cross section shape of the absorber tube on the performance of PTC system. The tested forms are the triangular, rectangular, and circular shapes. The geometries meshed by Gambit software are shown in Fig. 4. The details of the cross sections mesh are depicted in the same figure. For the three cases, we kept the same working fluid mass flow inlet and the same hydraulic diameter. The nonuniformity of the transversal and longitudinal meshes is shown in this figure.

Temperature contours of the absorber tube cross section for different tube shapes are presented in Fig. 6. The temperature evolutions in the flow direction and average temperatures of the heat transfer fluid circulating inside the tubes of different shapes are presented, respectively, in Fig. 5 and Table 3.

We notice that the circular cross section gives the highest outlet temperature (T_o). Indeed, the convective heat transfer from the absorber tube inner wall to the working fluid flowing in a circular tube was more important. This was proven by comparing the convective heat transfer coefficients (h_{cv}) of the different tube shapes summarized in Table 3.

Indeed, h_{cv} is higher for the absorber circular shape tube when compared to the rectangular or triangular cross section-shaped tubes.

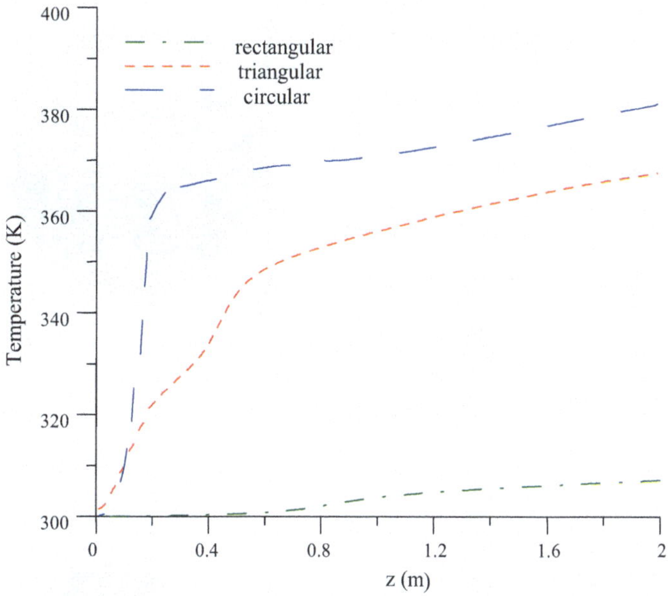

Fig. 5 Temperature evolution in the direction of flow for different cross section shapes

The main application of the PTC systems is to generate electricity. Indeed, solar electricity generation can be achieved by coupling PTC systems to a Rankine thermodynamic cycle. For this type of application, circular cross section tube configuration was chosen because it was the most effective. But, for other applications where electricity production is directly generated by photovoltaic cells placed on the lower periphery of the absorber tube, it is better to select a rectangular- or triangular-shaped tube in order to place the maximum of photovoltaic cells in the lower half of the tube surface receiving concentrated rays even if they do not give the highest temperature.

4 Conclusion

The thermal performance of the parabolic trough collector system was investigated. The first step consisted in determining the concentrated solar heat flux distribution by SOLTRACE software on the tube face exposed to the reflector. The receiver tube other side was subjected to direct sunlight transmitted through the glass envelope tube.

These heat flux densities were used in the second step as boundary conditions at the receiver tube wall during the numerical simulation using the FLUENT code. Various cross section shapes of the absorber tube were tested and compared in order to get the optimum shape for a better thermal performance.

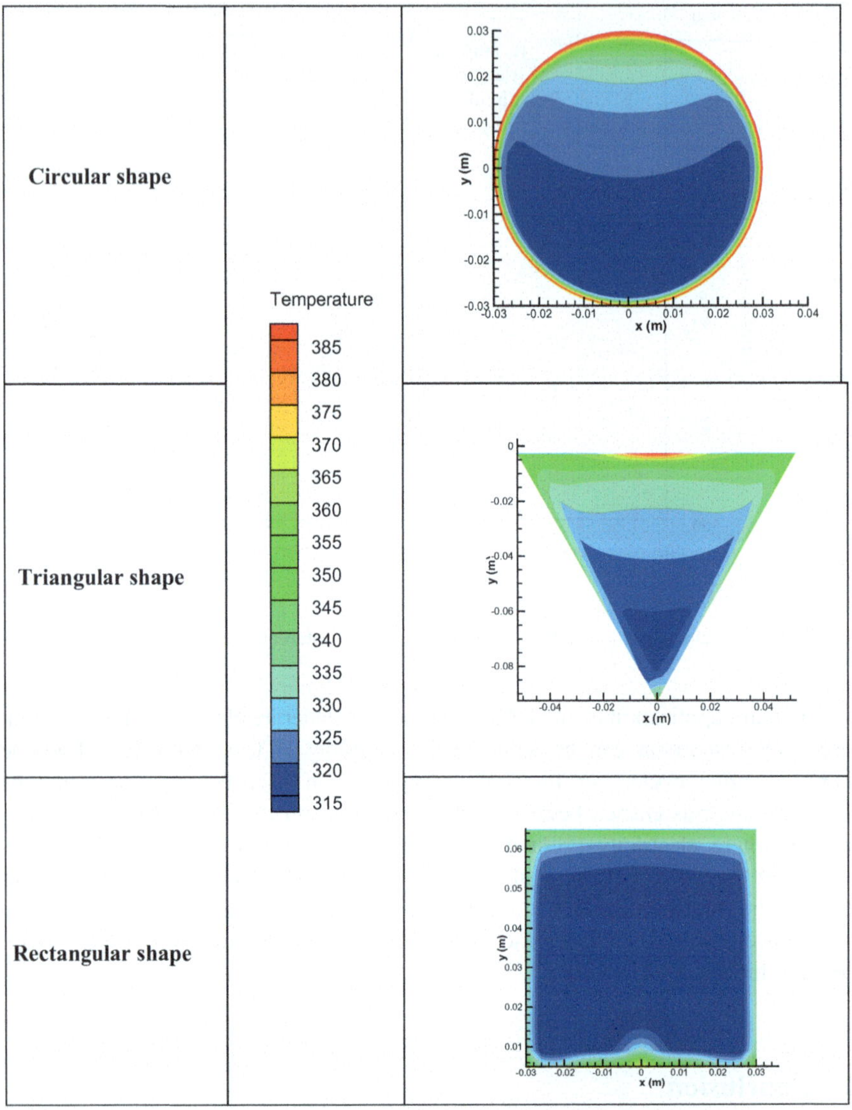

Fig. 6 Temperature contours in the outlet section for different cross section shapes

Table 3 T_o and h_{cv} for different cross section shapes of the absorber tube

	Circular	Triangular	Rectangular
h_{cv} (W/m^2 K)	132.077	90.817	84.362
T_o (K)	339.88	333.84	308.84

We arrived at the following conclusions:

(1) A good agreement was obtained between numerical results and Jeter's analytical data.
(2) The MCRT method integrated into SOLTRACE software was effective to estimate the solar flux for all PTC systems.
(3) The k-ε RNG turbulence model with near wall treatment and DO radiation model were able to predict the parameter characterizing the flow in a PTC system.
(4) The effect of the cross section shape of the absorber tube justified the cylindrical tubes choice in the solar thermodynamic installations.

References

Cheng ZD, He YL, Cui FQ (2012) Numerical study of heat transfer enhancement by unilateral longitudinal vortex generators inside parabolic trough solar receiver. Int J Heat Mass Transf 55 (21–22):5631–5641

Delatorre J, Baud G, Bezian JJ, Blanco S, Caliot C, Cornet JF, Coustet C, Dauchet J, El Hafi M, Eymet V, Fournier R, Gautrais J, Gourmel O, Joseph D, Meilhac N, Pajot A, Paulin M, Perez P, Piaud B, Roger M, Rolland J, Veynandt F, Weitz S (2014) Monte Carlo advances and concentrated solar applications. Sol Energy 103:653–681

Dongqiang L, Qiang L, Zhifeng W, Jian L, Jianbin L (2013) An experimental study of thermal characterization of parabolic trough receivers. Energy Convers Manag 69:107–115

FLUENT–Inc. (2005) FLUENT 6.2 User's Guide, FLUENT. Inc., FLUENT 6.3 Documentation

Gambit (2005) GAMBIT 2.3 User's Guide, FLUENT. Inc., GAMBIT 2.3 Documentation

Ghomrassi A, Mhiri H, Bournot P (2015) Numerical study and optimization of parabolic trough solar collector receiver tube. J Solar Energy Eng 137:051003-1

Jeter SM (1986) Calculation of the concentrated flux density distribution in parabolic trough collectors by a semi finite formulation. Sol Energy 37(5):335–346

Jeter SM (1987) Analytical determination of the optical performance of practical parabolic tough collectors from design data. Sol Energy 39(1):11–21

Lei D, Li Q, Wang Z, Li J (2013) An experimental study of thermal characterization of parabolic trough receivers. Energy Convers Manag 69:107–115

Mao QJ, Shuai Y, Yuan Y (2014) Study on radiation flux of the receiver with a parabolic solar concentrator system. Energy Convers Manag 84:1–6

Ouagued M, Khellaf A, Loukarfi L (2013) Estimation of the temperature, heat gain and heat loss by solar parabolic trough collector under Algerian climate using different thermal oils. Energy Convers Manag 75:191–201

Patankar SV, Spalding DB (1972) A calculation procedure for heat, mass and momentum transfer in three dimensional parabolic flows. Int J Heat Mass Transf 15(10):1787–1806

Ramchandra GP, Sudhir VP, Jyeshtharaj BJ (2014) Optimization of non-evacuated receiver of solar collector having non-uniform temperature distribution for minimum heat loss. Energy Convers Manag 85:70–84

Sargent and Lundy LLC Consulting Group (2003) Assessment of parabolic trough and power tower solar technology cost and performance forecasts. Report No. NREL/SR-550-34440

Usta Y (2010) Simulations of a large scale solar thermal power plant in turkey using concentrating parabolic trough collectors. thesis

Wang F, Shuai Y, Yuan Y, Yang G, Tan H (2010a) Effects of material selection on the thermal stresses of tube receiver under concentrated solar irradiation. Mater Desalin 33:284–291

Wang F, Shuai Y, Yuan Y, Yang G, Tan H (2010b) Thermal stress analysis of eccentric tube receiver using concentrated solar radiation. Solar Energy 84(10):1809–1815

Wang P, Liu DY, Xu C (2013) Numerical study of heat transfer enhancement in the receiver tube of direct steam generation with parabolic trough by inserting metal foams. Appl Energy 102:449–460

Windelin T, Dabos A (2013) SOLTRACE, a ray tracing code for complex solar optical systems. Technical Report No. NREL/TP-5500-59163

An OpenFOAM Solver for Forced Convection Heat Transfer Adopting Diagonally Implicit Runge–Kutta Schemes

Valerio D'Alessandro, Sergio Montelpare and Renato Ricci

1 Introduction

In this research, we developed CFD solvers for incompressible flows, based on open-source technology, adopting high-resolution time discretization schemes.

In theory, as suggested by many research papers, LES and DNS would not only require massively parallel computational resources but also accurate numerics. In particular, high-order time integration and spatial discretization methods would be preferred for ensuring minimal influence of numerical diffusion and dispersion on the flow. It is also very important to put in evidence that accurate numerics are typically adopted only in academic codes with a very limited dissemination to general public. Nevertheless open-source CFD codes provide suitable environments for the implementation (and spread) of low-dissipative algorithms needed for turbulence simulation.

The unstructured finite volume code OpenFOAM (Open-source Field Operation And Manipulation), Weller et al. (1998), was selected as a development environment in this work. It was released under the GNU Public License (GPL) and in the recent years, it has received great interest from the CFD community. Indeed, OpenFOAM object-oriented structure allows the users to implement own models and solvers into the baseline codes with relative effort, see for example,

V. D'Alessandro (✉)
Dipartimento di Ingegneria Industriale e Scienze Matematiche,
Università Politecnica delle Marche, Via Brecce Bianche, 60131 Ancona, Italy
e-mail: v.dalessandro@univpm.it

S. Montelpare · R. Ricci
Dipartimento di Ingegneria e Geologia, Università degli Studi "G. D'Annunzio" di
Chieti-Pescara, Viale Pindaro 42, 65127 Pescara, PE, Italy
e-mail: s.montelpare@unich.it

R. Ricci
e-mail: ricci@univpm.it

Z. Driss et al. (eds.), *CFD Techniques and Thermo-Mechanics Applications*,
https://doi.org/10.1007/978-3-319-70945-1_5

D'Alessandro et al. (2016), Buffo et al. (2016), Abdollahzadeh and Pascoa (2016), Fiorina et al. (2015), Silva and Lage (2011), Kassem et al. (2011), Favero et al. (2007). Hence, OpenFOAM features have been exploited in this research to devise the implementation of high-order explicit and implicit Runge–Kutta (RK) schemes for incompressible Navier–Stokes (INS) equations time integration.

Currently, the authors are aware of few Runge–Kutta methods implementations within the OpenFOAM code. As a matter of fact, Vuorinen et al. (2014) presented a seminal contribution treating the implementation of explicit Runge–Kutta based projection methods. On the other hand, Shen et al. (2010) focused their efforts on density-based solvers for all speeds adopting explicit Runge–Kutta techniques. Through this paper, we would like to contribute in the OpenFOAM-RK methods development, with emphasis on incompressible flows which were investigated previously only by Vuorinen et al. (2014). We have to note that the application of RK methods to the incompressible Navier–Stokes equations is not a trivial task because of the DAE nature of the discretized equations. Many CFD practitioners explicitly advance the velocity at each RK stage using standard techniques for ODE systems, thus the Poisson equation for the pressure is solved to guarantee a divergence-free velocity field. Anyhow, the impact of this projection-type approach on the temporal order of accuracy is not clear (Sanderse and Koren 2012). In many papers, the accuracy of the velocity is often darkly assumed to be unaffected by the DAE nature of INS. Besides, the temporal accuracy of the pressure is often not reported. Specifically in this work, an OpenFOAM solver based on Singly Diagonally Implicit Runge–Kutta (SDIRK) schemes is presented. Furthermore, as will be described in the next sections, an iterated PISO-like algorithm was used for handling pressure–velocity coupling within each RK stage; a specific procedure to preserve the theoretical order of accuracy was also adopted.

The effectiveness of the presented solver for incompressible Navier–Stokes equations was already addressed (D'Alessandro et al. 2016). Thus in this work, we extend our approach to forced convection heat transfer problems through the implementation of the energy equation.

For the previous reason, a suite of benchmark problems is considered in the following sections with the aim to demonstrate the reliability, the robustness, and the low dissipation of the proposed approach. Particular interest was devoted to the study of the bluff bodies forced convection heat transfer in order to provide a further contribution to this literature segment. The flow past these bodies is a typical complex fluid mechanics problem involving the interaction of: boundary layer, separating free-shear layer, and the wake. Moreover, under normal conditions, large flow separations are present around a bluff body thus creating a significant wake region downstream. Besides, these flows represent an important class of engineering applications and the induced unsteadiness might also be used to increase the heat transfer between the body and its surroundings. The previous cited flows are typically classified on the basis of: (i) number and arrangement of bluff bodies; (ii) cross-sectional shapes; (iii) approaching flow features; (iv) whether it is a free-stream or of a confined type.

Therefore, the forced convection heat transfer around both the circular cylinder and the square cylinder was studied in this research as well as the heat transfer from two circular cylinders in tandem configuration.

2 Governing Equations

The incompressible Navier–Stokes equations and energy equation read

$$\begin{aligned} &\nabla \cdot \mathbf{u} = 0 \\ &\frac{\partial \mathbf{u}}{\partial t} + \nabla \cdot (\mathbf{u} \otimes \mathbf{u}) = -\nabla p + \nabla \cdot (\nu \nabla \mathbf{u}) \\ &\frac{\partial \mathrm{T}}{\partial t} + \nabla \cdot (\mathbf{u}\mathrm{T}) = \nabla \cdot (\alpha \nabla \mathrm{T}) \end{aligned} \tag{1}$$

where **u** denotes the velocity vector, $p = P/\rho$ is the pressure divided by the density and T the temperature. Besides, ν is kinematic viscosity while α the thermal diffusivity.

2.1 Dimensionless Parameters

In this work, we presented the results in terms of the dimensionless parameters related to fluid dynamics and heat transfer: drag and lift coefficients, Strouhal number, and Nusselt number. The dimensionless drag and lift coefficients of the cylinder are defined by Eq. 2:

$$C_{\mathrm{D}} = \frac{2D}{\rho u_{\infty}^{2} A_{\mathrm{ref}}}, \quad C_{\mathrm{L}} = \frac{2L}{\rho u_{\infty}^{2} A_{\mathrm{ref}}}. \tag{2}$$

On the other hand, the Strouhal number (St) is defined according to Eq. 3

$$St = \frac{f L_{\mathrm{ref}}}{u_{\infty}} \tag{3}$$

where f is the vortex shedding frequency which is determined by spectral analysis of the time history of the fluctuating lift coefficient and L_{ref} the reference length.

The local Nusselt number Nu is calculated as

$$Nu = \frac{L_{\mathrm{ref}}}{T_{\mathrm{w}} - T_{\infty}} (\nabla T \cdot \mathbf{n})_{\mathrm{w}} \tag{4}$$

where T_w is the wall temperature, T_∞ the free-stream temperature, and λ the fluid thermal conductivity. The average Nusselt number, Nu_{avg}, can be obtained by integrating the Nu along the body surface.

3 Numerical Solution

In order to discretize Eq. 1, the computational domain is subdivided into N_c finite volumes and each volume is bounded by an arbitrary number of cell faces N_f. Thus, the Navier–Stokes and energy equations are rewritten in an integral form as regards each finite volume. The Gauss–Green divergence theorem is used to convert the volume integrals to surface integration over the mesh element boundary, so the semi-discrete equations read

$$\begin{aligned} &\sum_f^{N_f} \mathbf{U}_f \cdot \mathbf{S}_f = 0, \\ &\frac{d\mathbf{U}_P}{dt} = \underbrace{\frac{1}{\Omega_P}\left[-\sum_f^{N_f}(\mathbf{S}_f \cdot \mathbf{U}_f)\mathbf{U}_f + \sum_f^{N_f}\mathbf{S}_f \cdot (\nu\nabla\mathbf{U})_f\right]}_{\mathbf{F}_P} - (\nabla p)_P, \\ &\frac{dT_P}{dt} = \underbrace{\frac{1}{\Omega_P}\left[-\sum_f^{N_f}(\mathbf{S}_f \cdot \mathbf{U}_f)T_f + \sum_f^{N_f}\mathbf{S}_f \cdot (\alpha\nabla T)_f\right]}_{R_{T,P}}. \end{aligned} \tag{5}$$

Indeed, defining the residual vector, $\mathbf{R}_P$, as:

$$\mathbf{R}_{U,P}(\mathbf{U}_P, p, t) = \mathbf{F}_P - (\nabla p)_P \tag{6}$$

the discretized momentum equation showed in Eq. 5 can be expressed by

$$\frac{d\mathbf{U}_P}{dt} = \mathbf{R}_{U,P}(\mathbf{U}_P, p, t), \tag{7}$$

Similarly, the discretized energy equation is rewritten as

$$\frac{dT_P}{dt} = R_{T,P}(T_P, t) \tag{8}$$

where $R_{T,P}(T_P, t)$ is the energy equation residual defined in Eq. 5.

SDIRK methods allow solving at each stage a linear system with a rank equal to the spatial degrees of freedom. Such techniques, from an implementation point of view, are very attractive since each stage resembles a BDF1 scheme with a source term. Indeed, the SDIRK discretized governing equations can be expressed by

$$\frac{\mathbf{U}_{\mathrm{P}}^{(i)} - \mathbf{U}_{\mathrm{P}}^{(n)}}{\Delta t} = a_{ii}\mathbf{R}_{\mathrm{U,P}}^{(i)} + \sum_{j=1}^{i-1} a_{ij}\mathbf{R}_{\mathrm{U,P}}^{(j)}, \tag{9}$$

$$\frac{T_{\mathrm{P}}^{(i)} - T_{\mathrm{P}}^{(n)}}{\Delta t} = a_{ii}R_{\mathrm{T,P}}^{(i)} + \sum_{j=1}^{i-1} a_{ij}R_{T,\mathrm{P}}^{(j)}. \tag{10}$$

Furthermore in the next time step $\mathbf{U}_{\mathrm{P}}^{(n)}$ and $T_{\mathrm{P}}^{(n)}$ are updated by

$$\mathbf{U}_{P}^{(n+1)} = \mathbf{U}_{P}^{(n)} + \Delta t \sum_{j=1}^{s} b_j \mathbf{R}_{\mathrm{U,P}}^{(j)} \tag{11}$$

$$T_{P}^{(n+1)} = T_{\mathrm{P}}^{(n)} + \Delta t \sum_{j=1}^{s} b_j \mathrm{R}_{\mathrm{T,P}}^{(j)} \tag{12}$$

where s is the total number of RK stages.

It is also worth noting that in our work, for each implicit RK stage, we have used a PISO algorithm in order to handle the pressure–velocity coupling consisting of: (i) discretized momentum equation solution (predictor step); (ii) Poisson equation for pressure solution; (iii) cell-center velocity, face velocity, and mass-flux correction. As pointed out in the fundamental contribution published by Issa (1986), each additional PISO-iteration increases the order of accuracy of the splitting error by one. However, solving the elliptic pressure equation is the most time-consuming part for a PISO-loop, thus performing more than two or three iterations to lower the splitting error does not seem to offer a significant advantage. It is also very important to remark that the previous steps have been implemented in the spirit of standard PISO-based OpenFOAM solvers. This method gives an oscillation-free velocity in line with Rhie–Chow technique, even though there is no explicit Rhie–Chow procedure (Karrholm 2008).

We have also to note that the face residual vector $\mathbf{R}_{\mathrm{f}}^{(i)}$ is obtained by the discretized form of a SDIRK stage as in Kazemi-Kamyab et al. (2015). Kazemi-Kamyab et al. (2015) showed that this technique allows avoiding temporal order reduction suffered by several implementation techniques analyzed in the present literature. Moreover in this paper, we have used stiffly accurate SDIRK schemes having the following feature: $a_{sj} = b_j$. Hence, the solution of the last stage is equal to the solution of the new time level. The previous conditions allows avoiding the use of Eqs. 11 and 12.

In this research work, we have considered two different SDIRK techniques having, respectively, 3 (Alexander 1977) and 5 (Hairer 2010) stages. In particular, the first approach has a third order of accuracy (SDIRK 3-3) while the second the fourth order (SDIRK 5-4). As for the spatial discretization of the diffusive terms and pressure gradients, second-order accurate schemes were used. The convective terms were discretized using a central difference scheme, for momentum and energy equations.

Lastly, we have put in evidence that the previous solution algorithms were implemented using exclusively OpenFOAM basic classes. This element confirms that OpenFOAM is a very powerful tool to mimic typical PDE language used in CFD and in other scientific fields. Only for the Nusselt number evaluation did we have to rely, for simplicity purposes, on the "external" library `swak4Foam`.[1]

4 Results

In this section, we present solutions for different convective heat transfer problems. It should be noted that a Prandtl number, Pr, equal to 0.71 was selected for the reported cases.

Concerning the linear solvers, a preconditioned bi-conjugate gradient method (PBiCG) with DILU preconditioner was used to solve the discretized momentum and energy equations. On the other hand, a preconditioned conjugate gradient method (PCG) with a diagonal incomplete Cholesky preconditioner was adopted for the pressure. In particular, a local accuracy of 10^{-7} was fixed for the pressure; differently the systems for the velocity and the temperature were considered converged when the residuals reached the machine precision. Lastly, the PISO iterations, n_p, were set equal to 3 within each RK substage.

All the numerical solutions have been computed on a very small Linux Cluster, built within our research group, with five nodes consisting of Intel Core2-Quad Q9550 for a total of 20 cores operating at 2.8 GHz. It can be considered a trashware computing solution however it provides sufficient performance to compute the low Reynolds number problems considered in this work.

4.1 Taylor–Green Vortex

The first flow problem here considered is the Taylor–Green vortex. It is a classical flow that is widely used for testing the ability of a numerical method in solving the smallest scales in turbulent flows. In this test, the velocity and pressure fields are initialized with

[1] `swak4Foam` library website: https://openfoamwiki.net/index.php/Contrib/swak4Foam.

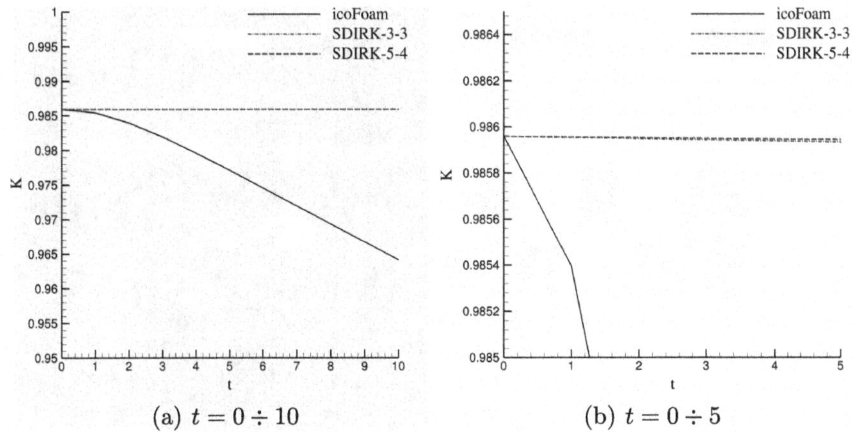

Fig. 1 Total kinetic energy of inviscid the Taylor–Green vortex

$$\begin{aligned} \mathbf{u}(x,y,t) &= [-\sin(\pi x)\cos(\pi y)\mathbf{i} + \cos(\pi x)\sin(\pi y)\mathbf{j}]e^{-2\pi^2 t/Re} \\ p(x,y,t) &= \frac{1}{4}[\cos(2\pi x) + \cos(2\pi y)]e^{-4\pi^2 t/Re}. \end{aligned} \tag{13}$$

Here, we use this flow in a different way on a square computational domain with: $\Omega = [-1, 1]$. Indeed, the aim of this test is to demonstrate that the RK schemes, described in Sect. 2, introduce a very small amount of numerical dissipation. The flow inviscid limit, $\nu = 0$, with periodic boundary conditions was studied, since in these conditions, the total kinetic energy of the vortex is conserved: $\mathrm{d}K/\mathrm{d}t = 0$. Dissipative numerical solutions, see for example Fig. 1, show that $\mathrm{d}K/\mathrm{d}t < 0$, thus this test case is very suitable to evaluate the schemes performance. From Fig. 1, it is very easy to see that standard OF solver for incompressible Navier–Stokes equations, i.e, `icoFoam`, is very dissipative, while the RK-based solvers presented here have low-dissipation properties, as expected. Hence, the main purpose of this work was reached.

4.2 *Circular Cylinder*

The forced convection heat transfer problem of a single isothermal cylinder at $Re = 100$ and 200, see Fig. 2 for a solution representation, was considered as a first validation case. A computational grid with O-type topology having $N_c = 6 \times 10^4$, was used. The far field boundary was placed at 25 times the cylinder diameter and the height of the first cell next to the wall was fixed equal to $5 \times 10^{-3}D$. The time step size Δt was selected equal to $10^{-2}D/u_\infty$.

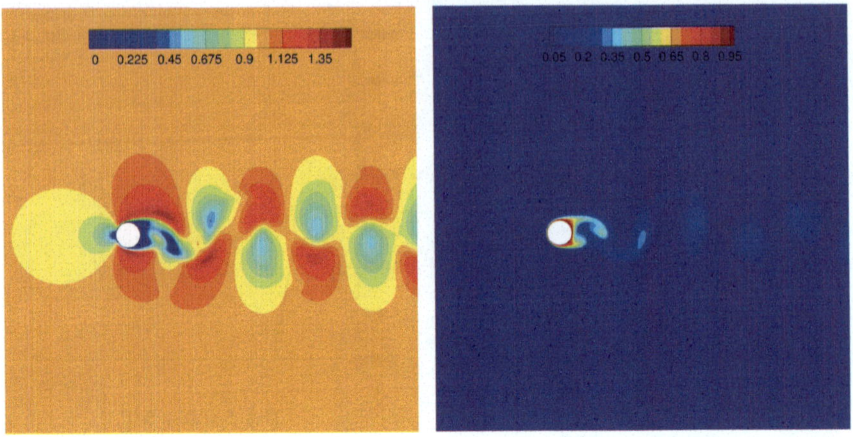

(a) Velocity magnitude (b) Temperature

Fig. 2 Circular cylinder contours plots, $Re = 200$ at $t = 300D/u_\infty$

The drag and lift coefficients and Strouhal number were computed and comparatively tabulated with literature numerical and experimental data in Tables 1 and 2. Comparison of the obtained results with literature data shows a good agreement.

The average Nusselt numbers of a single isothermal cylinder in cross-flow at $Re = 100$ and 200 are summarized in Table 3. Also in this case, the results available in literature are reported as well as Churchill and Bernestein (1977) correlation and predictions. It can be seen that the computed Nusselt numbers are in good agreement with literature values. Moreover, the results obtained using the SDIRK 3-3 scheme and the SDIRK 5-4 are almost indistinguishable; hence in the following, for the sake of compactness, only SDIRK 3-3 scheme is adopted. We have to note that for the case at $Re = 100$, a sensitivity study to the number of PISO-corrector within each RK stage was also performed. We have increased the correctors up to 5 obtaining almost indistinguishable results (the data are not

Table 1 Force coefficients and Strouhal number at $Re = 100$

	C_L	C_D	St
SDIRK 3-3	±0.333	1.340 ± 0.010	0.165
SDIRK 5-4	±0.332	1.340 ± 0.010	0.165
Mahir and Altac (2008)	±0.343	1.368 ± 0.029	0.172
Braza et al. (1986)	±0.25	1.364 ± 0.015	0.160
Ding et al. (2007)	±0.287	1.356 ± 0.010	0.166
Liu et al. (1998)	±0.339	1.35 ± 0.012	0.164
Norberg (2003), (Exp.)	±0.18 to ±0.54	–	0.168
Williamson (1991), (Exp.)	–	–	0.164
Harimi and Saghafian (2012)	±0.306	1.344 ± 0.007	0.165

Table 2 Force coefficients and Strouhal number at $Re = 200$

	C_L	C_D	St
SDIRK 3-3	± 0.690	1.3347 ± 0.046	0.195
SDIRK 5-4	± 0.690	1.3357 ± 0.046	0.195
Mahir and Altac (2008)	± 0.698	1.376 ± 0.048	0.192
Braza et al. (1986)	± 0.75	1.4 ± 0.05	0.200
Ding et al. (2007)	± 0.659	1.348 ± 0.050	0.196
Liu et al. (1998)	± 0.69	1.31 ± 0.049	0.192
Norberg (2003), (Exp.)	± 0.35 to ± 0.70	–	0.18 to 0.197
Williamson (1991), (Exp.)	–	–	0.196
Harimi and Saghafian (2012)	0.656	1.337 ± 0.04	0.194

Table 3 Average Nusselt number at $Re = 100$ and 200, $Pr = 0.71$

	$Re = 100$	$Re = 200$
SDIRK 3-3	5.153 ± 0.0022	7.452 ± 0.0270
SDIRK 5-4	5.153 ± 0.0021	7.452 ± 0.0270
Mahir and Altac (2008)	5.179 ± 0.003	7.474 ± 0.028
Harimi and Saghafian (2012)	5.0613	7.1079
Mahfouz and Badr (2000)	5.31	6.99
Patnana et al. (2010)	5.154	–
Churchill and Bernstein (1977)	5.156	7.188

reported in the Tables for compactness); hence our base setting ($n_p = 3$) was retained in the following computations.

A further important element was addressed using this test case: the computational cost. SDIRK 3-3 scheme requires 352 s using 4 CPU cores (on the adopted computational facility) in order to compute a complete vortex shedding cycle at $Re = 100$, with a memory allocation of about 357 MB. The SDIRK 5-4 scheme requires a wall clock time of 593 s, allocating also 393 MB of RAM memory. On the other hand, the standard `icoFoam` solver, equipped with energy equation, employs 161 s for the computation of shedding cycle using 192 MB of memory. It is very easy to see that our RK-based solvers do not require a great amount of memory, this is due to the segregated approach used within each RK stage. So the main drawback related to the solver presented here is represented by the increase in CPU time. However, we have to point out that improving the parallel scalability of the OpenFOAM linear solvers could significantly reduce the previous problem.

4.3 Square Cylinder

In this subsection, we present numerical computations of the unsteady flow over a square cylinder using the classical uniform temperature boundary condition on the cylinder wall. It is interesting to note that the heat and fluid flow around a circular and a square cylinder share the main features of a typical bluff body flow. However it is important to put in evidence that the separation mechanism, shedding frequency, and aerodynamic forces depend differently from Reynolds number. For the square cylinder, the separation points are fixed, differently for the circular shape the separation point varies with *Re* number. Furthermore, the width of the wake immediately behind the body is at least one diameter for the square shape, while it is less than half a diameter for the circular cylinder.

In this work, we have considered several *Re* numbers, ranging from 50 to 160, see Fig. 3 for a solution representation. In this *Re* range, an unsteady flow is expected so this configuration is particularly attractive for the validation purposes of our numerical solution technique.

For these computations, we have adopted the same domain topology and dimensions proposed by Sharma and Eswaran (2004). Moreover, the first cell height next to the wall was fixed equal to $5 \times 10^{-3} D$, while the number of cells used to discretize the domain was $N_c = 10^5$. On the other hand, the time step size was $2 \times 10^{-3} D/u_\infty$ for all the different flow problems considered here.

In Fig. 4a, the variation of the drag coefficient versus *Re* number is plotted along the literature experimental and numerical data, (Sharma and Eswaran 2004; Sohankar et al. 1995; Franke et al. 1990; Shimizu and Tanida 1978). The trend of variation of the C_D coefficient, obtained in the present work, is very close to Sharma

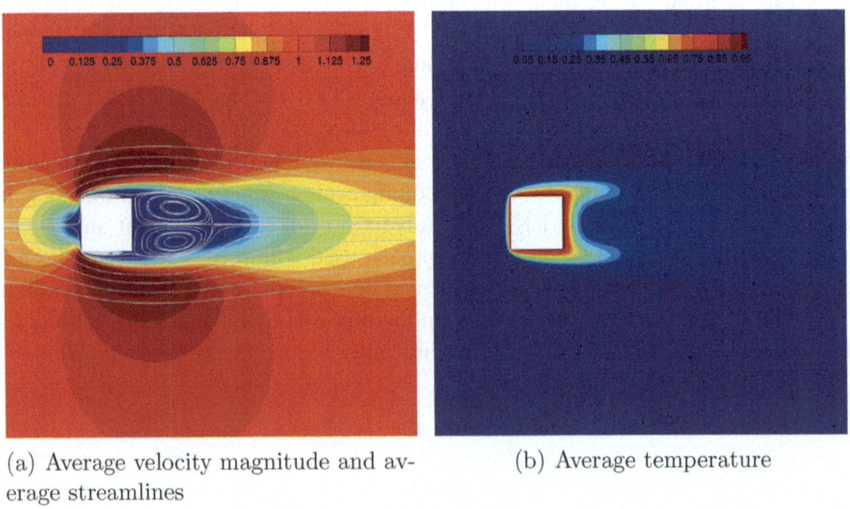

(a) Average velocity magnitude and average streamlines

(b) Average temperature

Fig. 3 Square cylinder contours plots, $Re = 150$

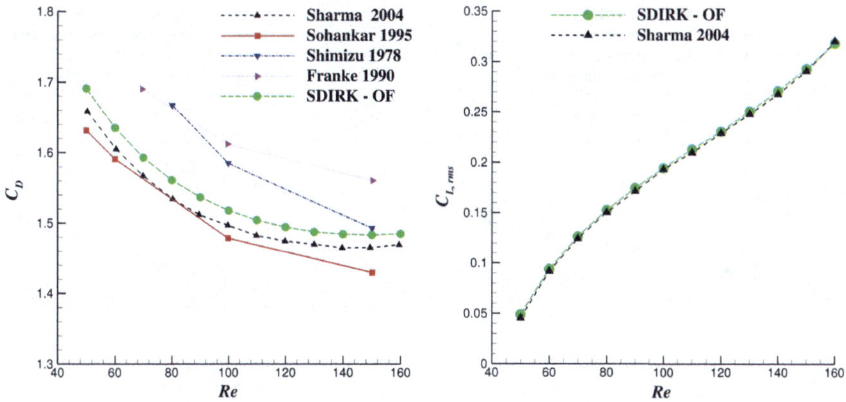

(a) Drag coefficient (b) Lift Coefficient RMS

Fig. 4 Square cylinder. Force coefficients at various *Re* numbers

and Eswaran (2004), as well as to Sohankar et al. (1995). In addition, the variation of rms of C_L with *Re* numbers is shown in Fig. 4b. It is worth noting that $C_{L,rms}$ gives a measure of the amplitude of the unsteady cylinder wake oscillations. In this specific case, our results are in very good agreement with Sharma and Eswaran (2004).

Furthermore, Fig. 5a shows the mean recirculating zone length at different *Re*. This figure shows the monotonic decrease in the mean wake length with increasing the *Re* number. Note that this trend is also consistent with the 2-D flow over the circular cylinder. The previous figure shows that the present computations agree closely with the results of Sharma and Eswaran (2004). Figure 5b reports the

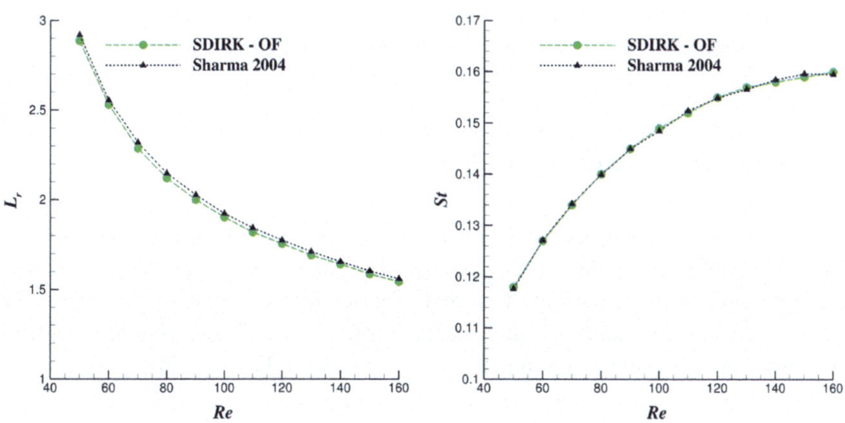

(a) Average recirculating length (b) Strouhal number

Fig. 5 Square cylinder. Flow parameters at various *Re*

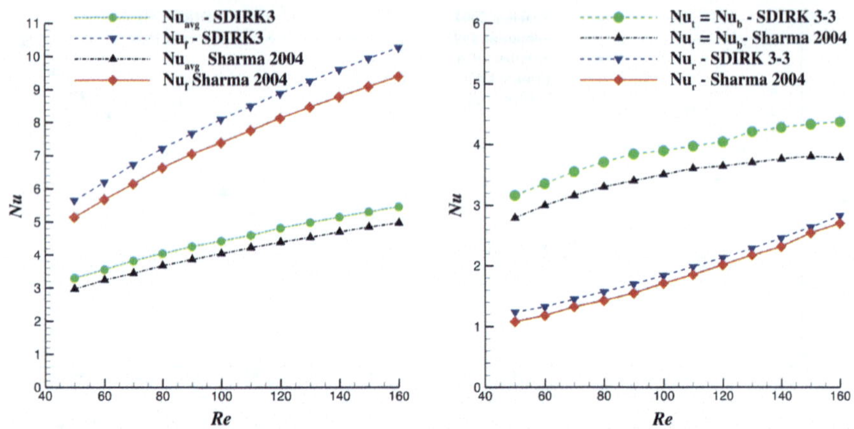

(a) Average Nu and front Nu (b) top and bottom Nu and rear Nu

Fig. 6 Square cylinder. Nusselt number at various *Re*

Strouhal versus Reynolds number variation obtained in the present simulations along with the corresponding results of Sharma and Eswaran (2004). Also in this case, a very good agreement with the literature data is obtained.

Figure 6 shows the variation of the average cylinder Nusselt number, as well as that of each surface separately, versus the *Re* number.

It is very easy to see that the front surface consistently exhibits the highest Nu, the top and the bottom surfaces values are intermediate and the rear surface has the lowest average *Nu* number value. The front *Nu*, Nu_f, is slightly over-estimated if compared with the reference data. The same condition is evidenced for the rear *Nu*, Nu_r. Differently, a good agreement is obtained for the top and bottom surfaces. Globally, our results evidence the same trend of the Sharma and Eswaran (2004) results, and just like (Sharma and Eswaran 2004), we have obtained an average over-estimation error of about 8.5% for the considered *Re* numbers.

4.4 Tandem Circular Cylinders

The last benchmark considered is related to the convective heat transfer of two circular cylinders in tandem configuration. The studies involving this flow configuration have been the subject of several experimental and numerical works in the last two decades because of the practical relevance of this problem and its fluid-dynamic complexity. Examples of its various applications in practical engineering areas include: offshore platforms, transmission cables, cooling towers, heat exchanger tubes, chimney stacks, and marine riser.

In this work, we have considered a $Re = 100$ and $Re = 200$ with a center-to-center (L/D) distance keept to 5 and a fixed temperature condition for the

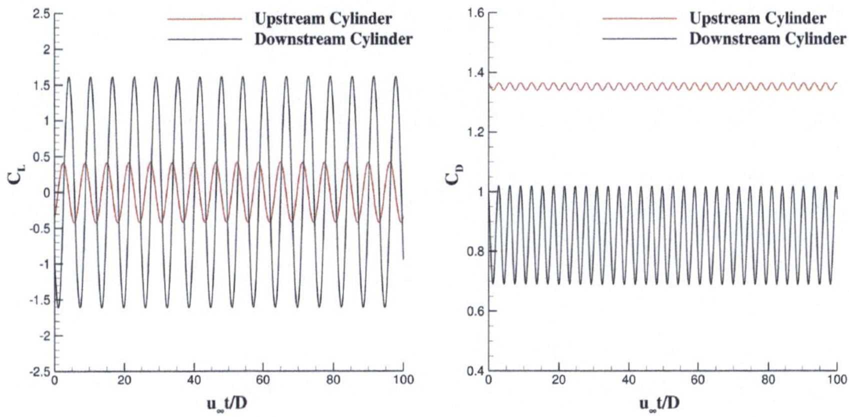

(a) Lift coefficient (b) Drag coefficient

Fig. 7 Tandem cylinders, force coefficients time histories, $Re = 100$

cylinders walls. For this case, vortex shedding is expected behind both the cylinders (Mahir and Altac 2008), as can be easily observed in Fig. 7. Regarding the domain shape and extension, we have adopted the same features proposed by Mahir and Altac (2008). We have to remark that in this work, we have used a structured grid with $N_c = 1.6 \times 10^5$, see Fig. 8a, while Mahir and Altac built a fully unstructured (triangular elements) grid. A first cell height, near the cylinders walls, was fixed to $5 \times 10^{-3} D$ with a value of time step fixed to $10^{-3} D/u_\infty$. Finally in this specific case, we have obtained stable computation only fixing 4 PISO iterations within each SDIRK substage.

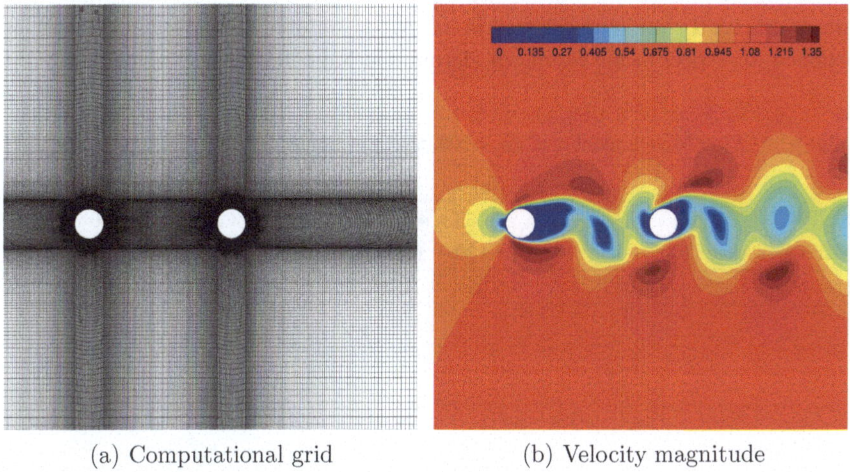

(a) Computational grid (b) Velocity magnitude

Fig. 8 Tandem cylinders, $Re = 100$

Table 4 Force coefficients and Strouhal number at $Re = 100$, tandem cylinders

	$C_{L,1}$	$C_{D,1}$	$C_{L,2}$	$C_{D,2}$	$St_{,1}$
SDIRK 3-3	± 0.417	$1.3535 \pm 1.23 \times 10^{-2}$	± 1.615	$0.853 \pm 1.65 \times 10^{-1}$	0.160
Mahir and Altac (2008)	± 0.437	$1.369 \pm 1.3 \times 10^{-2}$	± 1.617	$0.874 \pm 1.65 \times 10^{-1}$	0.161
Harimi, and Saghafian, (Harimi and Saghafian 2012)	–	~ 1.3	–	~ 0.8	0.157

Table 5 Force coefficients and Strouhal number at $Re = 200$, tandem cylinders

	$C_{L,1}$	$C_{D,1}$	$C_{L,2}$	$C_{D,2}$	$St_{,1}$
SDIRK 3-3	± 0.712	$1.3180 \pm 5.54 \times 10^{-2}$	± 1.625	$0.490 \pm 1.82 \times 10^{-1}$	0.187
Mahir and Altac (2008)	± 0.731	$1.327 \pm 5.5 \times 10^{-2}$	± 1.569	$0.455 \pm 1.6 \times 10^{-1}$	0.186
Harimi and Saghafian (2012)	–	~ 1.3	–	~ 0.6	0.182

The average force coefficients, C_L and C_D, and their oscillations are collected in Tables 4 and 5. As it can be observed, the obtained coefficients agree well with those of Mahir and Altac (2008) and Harimi and Saghafian (2012).

It is worth emphasizing that separation and vortex shedding from the cylinder surface into its wake can result in the variation of local Nusselt number, thus also Nu_{avg} oscillates periodically in time. Tables 6 and 7 report the comparison between our Nu_{avg} and the literature references, i.e, Mahir and Altac (2008), Harimi and Saghafian (2012). Note that for this specific problem, the average Nu of the upstream cylinder approaches to that of a single cylinder. It is also particularly interesting to observe that our results are in very good agreement with Mahir and Altac, differently an over-estimation of Nu_{avg} on the downstream cylinder is obtained in respect to the results of Harimi and Saghafian.

Table 6 Average Nusselt number at $Re = 100$, tandem cylinders

	$Nu_{,1}$	$Nu_{,2}$
SDIRK 3-3	$5.185 \pm 4.5 \times 10^{-3}$	$4.27 \pm 4.12 \times 10^{-1}$
Mahir and Altac (2008)	$5.180 \pm 5.0 \times 10^{-3}$	$4.28 \pm 4.0 \times 10^{-1}$
Harimi and Saghafian (2012)	5.028	3.264

Table 7 Average Nusselt number at $Re = 200$, tandem cylinders

	$Nu_{,1}$	$Nu_{,2}$
SDIRK 3-3	$7.44 \pm 3.15 \times 10^{-2}$	$5.977 \pm 5.04 \times 10^{-1}$
Mahir and Altac (2008)	$7.43 \pm 3.4 \times 10^{-2}$	$5.96 \pm 4.7 \times 10^{-1}$
Harimi and Saghafian (2012)	7.027	4.427

5 Conclusions

In this work, we have presented OpenFOAM solvers for forced convection heat transfer problems based on high-order diagonally implicit RK schemes for time integration. The presented solvers adopt a PISO-iterated procedure with RK-substage in order to handle pressure–velocity coupling. Moreover, the PISO-loop is implemented in the spirit of standard OpenFOAM solvers in order to obtain an oscillation-free velocity.

The effectiveness, reliability, and robustness of the proposed approach were evaluated computing several standard heat transfer benchmarks. This is a point of interest since we cannot find papers, in the available literature, about the computation of forced convection heat transfer problems using SDIRK schemes based on a PISO-like procedure. Our interest was mainly devoted to the study of the bluff bodies forced convection heat transfer, in order to provide a further contribution to this complex literature segment. Note that these flows represent an important class of engineering applications. Moreover, the low-dissipation properties of the adopted schemes were assessed considering the Taylor–Green vortex problem.

The achieved results prove, in our opinion, that our OpenFOAM solvers are extremely well suited to efficiently and accurately address the DNS and LES of turbulent flows involving heat transfer. Currently, the presented solvers have been extended also to large eddy simulation. Challenging computations related to jet impingement heat transfer are under investigation, and the obtained results will be presented in future papers.

References

Abdollahzadeh M, Pascoa JC, Oliveira PJ (2016) Implementation of the classical plasma-fluid model for simulation of dielectric barrier discharge (DBD) actuators in OpenFOAM. Comput Fluids 128:77–90

Alexander R (1977) Diagonally implicit Runge-Kutta methods for stiff O.D.E.'s. SIAM J Numer Anal 14:1006–1021

Braza M, Chassaing P, Ha Minh H (1986) Numerical study and physical analysis of the pressure and velocity fields in the near wake of a circular cylinder. J Fluid Mech 165:79–130

Buffo A, Vanni M, Marchisio DL (2016) On the implementation of moment transport equations in OpenFOAM: Boundedness and realizability. Int J Multiph Flow 85:223–235

Churchill SW, Bernstein MJ (1977) A correlating equation for forced convection from gases and liquids to a circular cylinder in cross flow. J Heat Transf 99:300–306

D'Alessandro V, Zoppi A, Binci L, Ricci R (2016) Development of OpenFOAM solvers for incompressible Navier-Stokes equations based on high-order Runge-Kutta schemes. Advances in Fluid Mechanics XI, Ancona-Italy

D'Alessandro V, Montelpare S, Ricci R (2016b) Detached-eddy simulations of the flow over a cylinder at Re = 3900 using OpenFOAM. Comput Fluids 136:152–169

Ding H, Shu C, Yeo YO, Xu D (2007) Numerical simulation of flows around two circular cylinders by mesh-free least square-based finite difference methods. Int J Numer Methods Fluids 53:305–322

Favero JL, Secchi AR, Cardozo NSM, Jasak H (2007) Viscoelastic flow simulation: development of a methodology of analysis using the software OpenFOAM and differential constitutive equations. Comput Aided Chem Eng 27:915–920

Fiorina C, Clifford I, Aufiero M, Mikityuk K (2015) GeN-Foam: a novel OpenFOAM based multi-physics solver for 2D/3D transient analysis of nuclear reactors. Nucl Eng Des 294:24–37

Franke R, Rodi W, Schonug B (1990) Numerical calculation of laminar bortex-shedding flow past cylinders. J Wind Eng Ind Aerodyn 35:237–257

Hairer E, Wanner G (2010) Solving ordinary differential equation II, stiff and differential-algebraic problems. Springer, 1996

Harimi I, Saghafian M (2012) Numerical simulation of fluid flow and forced convection heat transfer from tandem circular cylinders using overset grid method. J Fluids Struct 28:309–327

Issa RI (1986) Solution of the implicitly discretised fluid flow equations by operator-splitting. J Comput Phys 62:40–65

Karrholm FP (2008) Numerical modelling of diesel spray injection, turbulence interaction and combustion. Ph.D. thesis, Chalmers University of Technology

Kassem HI, Saqr KM, Aly HS, Sies MM, Wahid MA (2011) Implementation of the eddy dissipation model of turbulent non-premixed combustion in OpenFOAM. Int Commun Heat Mass Transfer 38:363–367

Kazemi-Kamyab V, van Zuijlen A, Bijl H (2015) Analysis and application of high order implicit Runge-Kutta schemes to collocated finite volume discretization of the incompressible Navier-Stokes equations. Comput Fluids 109:107–115

Liu C, Zheng X, Sung CH (1998) Preconditioned multigrid methods for unsteady incompressible flows. J Comput Phys 139:35–37

Mahfouz FM, Badr HM (2000) Forced convection from a rotationally oscillating cylinder placed in a uniform stream. Int J Heat Mass Transf 43:3093–3104

Mahir N, Altac Z (2008) Numerical investigation of convective heat transfer in unsteady flow past two cylinders in tandem arrangements. Int J Heat Fluid Flow 29:1309–1318

Norberg C (2003) Fluctuating lift on a circular cylinder: review and new measurements. J Fluids Struct 17:57–96

Patnana VK, Bharti RP, Chabra RP (2010) Two-dimensional unsteady forced convection heat transfer in power-law fluids from a cylinder. Int J Heat Mass Transf 53:4152–4167

Sanderse B, Koren B (2012) Accuracy analysis of explicit Runge-Kutta methods applied to the incompressible Navier-Stokes equations. J Comput Phys 231:3041–3063

Sharma A, Eswaran V (2004) Heat and fluid flow across a square cylinder in the two-dimensional laminar flow regime. Numer Heat Transf Part A 45:247–269

Shen C, Sun F, Xia X (2010) Implementation of density-based solver for all speeds in the framework of OpenFOAM. Comput Phys Commun 185:2730–2741

Shimizu Y, Tanida Y (1978) Fluid forces acting on cylinders of rectangular cross section. Trans Jpn Soc Mech Eng B 44:2699–2706

Silva LFLR, Lage PLC (2011) Development and implementation of a polydispersed multiphase flow model in OpenFOAM. Comput Chem Eng 35:2653–2666

Sohankar A, Norberg C, Davidson L (1995) Numerical simulations of unsteady flow around a square two-dimensional cylinder. In: Proceedings of 12th Australian fluid mechanics conference, pp 517–530

Vuorinen V, Keskinen JP, Duwig C, Boersma BJ (2014) On the implementation of low-dissipative Runge-Kutta projection methods for time dependent flows using OpenFOAM. Comput Fluids 93:153–163

Weller HG, Tabor G, Jasak H, Fureby C (1998) A tensorial approach to computational continuum mechanics using object-oriented techniques. Comput Phys 12:620–631

Williamson CHK (1991) 2-D and 3-D aspects of the wake of a cylinder and their relation to wake computations. In: Anderson CR, Greengard C (eds) Vortex dynamics and vortex methods, vol 28. American Mathematical Society. Lectures in Applied Mathematics, pp 719–751

Multigrid and Preconditioning Techniques in CFD Applications

Konstantin Volkov

1 Introduction

Finite volume method and high-resolution schemes are applied to discretization of Euler and Navier–Stokes equations on unstructured meshes.

1.1 Preconditioning

Fluid flows are described by a system of equations consisting of the continuity equation, momentum equation, energy equation, and an equation of state (Volkov 2008). The momentum equation has various forms in the inviscid and viscous models (Euler equations for inviscid flows and the Navier–Stokes equation for viscous flows) (Wesseling 2000).

The artificial compressibility (pseudo-compressibility) method (Chorin 1967), in which the time derivative of pressure is introduced into the continuity equation; the projection method (Chorin 1968), which is based on splitting over physical processes; and pressure projection methods (Patankar and Spalding 1972) are widely used for integrating the incompressible flow equations. A common feature of pressure projection methods is that the difference scheme is formulated in terms of increments of unknown functions, and the Poisson equation is solved for the pressure correction term at every time step. In the case of implicit difference schemes, the alternating direction implicit (ADI) method (Briley and McDonald 1977),

K. Volkov (✉)
ITMO University, St Petersburg 197101, Russia
e-mail: k.volkov@kingston.ac.uk

K. Volkov
Faculty of Science, Engineering and Computing,
Kingston University, London SW15 3DW, UK

Z. Driss et al. (eds.), *CFD Techniques and Thermo-Mechanics Applications*,
https://doi.org/10.1007/978-3-319-70945-1_6

relaxation-type methods (MacCormack 1985), the lower-upper symmetric Gauss–Seidel (LU-SGS) method (Yoon and Kwak 1989), and others are widely used. An overview and comparison of various approaches is given in Kwak et al. (2011).

Numerical methods for compressible Navier–Stokes equations that perform well at moderately subsonic and supersonic flow velocities become low effective or unsuitable when applied to flows at low Mach numbers ($M < 0.2$) (Wesseling 2000), which is manifested by slower convergence of time marching to a steady state and by the loss of accuracy of the resulting steady-state solutions (Choi and Merkle 1985, 1993; Merkle and Choi 1988; Volpe 1993). The slower convergence of time marching is explained by the fact that the stiffness of the compressible Euler and Navier–Stokes equations increases as $M \to 0$ (this feature is exhibited at the differential level). The stiffness is characterized by the ratio of the maximum to minimum eigenvalues of the Jacobian (the ratio of the maximum to minimum propagation speeds of perturbations). The integration time step is determined by the speed of the fastest wave (acoustic waves, $\lambda = |u + c|$), while the time required for reaching a steady state depends on the speed of the slowest wave (convective waves, $\lambda = |u|$). In viscous problems and turbulent flow computations on stretched meshes in boundary layers, the time step is restricted by the acoustic solution modes and by the mesh size in the normal direction to the wall (Pierce and Giles 1996, 1997).

The numerical simulation of flows at low Mach numbers is based on the incompressible Euler or Navier–Stokes equations with the use of suitable methods. For $M < 0.3$, the incompressible fluid model provides a fairly accurate approximation with an error of about 5%. The full Euler or Navier–Stokes equations are required for simulating high-speed flows with extended low-speed subregions (Volkov 2011) (e.g., flows with recirculation zones, internal flows in diffusers with a subsonic inlet velocity) and low-speed flows with density and temperature variations caused by heat supply (for example, free convective flows).

The transition to the limit form (as $M \to 0$) of the Navier–Stokes equations for hypersonic non-isothermal viscous flows makes it possible to partially eliminate the difficulties arising from computing these flows relying on the full Navier–Stokes equations (Weiss and Smith 1995).

A popular method for eliminating the computational difficulties arising as $M \to 0$ is related to various preconditioning techniques, which are aimed at leveling the orders of the eigenvalues of the Jacobian for $M < 1$ (Turkel et al. 1993, 1996, 1997; Turkel 1997; Darmofal and Schmid 1996; Guillard and Viozat 1999; Weiss et al. 1999). At the differential level, preconditioning modifies the terms involving time derivatives in the momentum equations. In a steady state, the solution of the modified (preconditioned) equations coincides with that of the original system of equations. An unsteady solution is found by applying dual time-stepping procedure (Lee 2007).

Preconditioning is also widely used to accelerate the convergence of the iterative methods as applied to systems of difference equations generated by finite difference or finite volume discretizations of the Euler and Navier–Stokes equations.

Preconditioning makes it possible to eliminate the stiffness of the original system and accelerate the convergence of time marching to a steady state (Choi and Merkle

1985, 1993; Merkle and Choi 1988; Volpe 1993). Additionally, subsonic flows can be computed more accurately by applying a modified discretization of convective fluxes in the preconditioned equations (Shuen et al. 1993; Housman et al. 2009). In the general case, preconditioning changes the form of the underlying equations and the properties of difference schemes because it introduces an artificial viscosity. The accuracy of preconditioned difference schemes degrades with increasing the Mach number. Theoretical issues related to the preconditioning of the Euler and Navier–Stokes equations at low Mach numbers are discussed in Kwak et al. (2011), Turkel et al. (1997), Venkateswaran and Merkle (1999), and Darmofal et al. (2000), while various preconditioning approaches are compared in Colin et al. (2011).

The methods developed in Choi and Merkle (1985, 1993), Merkle and Choi (1988) and those proposed in Turkel et al. (1993, 1996), Turkel (1997) are widely used in computational practice. The field of application of the method from Choi and Merkle (1985, 1993), Merkle and Choi (1988) is restricted to central difference schemes (CDS), which perform well at $M < 1$, but become dissipative in supersonic flow simulation. The method of Turkel et al. (1993, 1996), Turkel (1997) can easily be applied to upwind difference schemes (UDS) and has been widely used in external CFD simulations (Pierce and Giles 1996, 1997). The local preconditioning of the Euler and Navier–Stokes equations was applied in Moinier and Giles (2001), Volkov (2009). In this case, the transition to modified equations depends on the local Mach number (external flows) or the local pressure field (internal flows). In many cases, preconditioning methods are combined with other convergence acceleration methods (Volkov 2011), such as residual smoothing and multigrid methods.

While numerical computations are usually based on equations written in conservative variables, the preconditioning matrix is constructed using physical variables, which simplify the construction procedure (Turkel et al. 1997). Entropy (symmetrized variables) is used as a dependent variable in Turkel et al. (1997), while temperature (physical variables) is applied in Choi and Merkle (1985, 1993), Merkle and Choi (1988), and Weiss and Smith (1995). A preconditioning matrix that modifies only the energy equation is used in Choi and Merkle (1985), Merkle and Choi (1988). A method intended for simulating viscous flows is presented in Choi and Merkle (1993). The preconditioning procedure in van Leer et al. (1991) is designed so as to optimize the propagation velocities of waves in the entire range of Mach numbers (optimal condition number).

For the Euler and Navier–Stokes equations, a preconditioning method is developed. It makes it possible to construct a universal numerical procedure for computing inviscid and viscous compressible gas flows in a wide range of Mach numbers (from essentially subsonic to transonic and supersonic flows). The preconditioning matrix is constructed by applying the approach proposed in Weiss and Smith (1995), which was implemented in the one-dimensional case in Wesseling (2000). This approach relies on physical variables (one of which is temperature). Its features include a specific form of writing fluxes, the computation of a dissipative term in the course of finding the fluxes through control volume faces, and a specific representation of matrices in the diagonalization of the inviscid flux Jacobian of the preconditioned system. The dissipative term in the difference scheme for flux computation is written in a compact form.

1.2 Free Convective Flows

The equations of continuity and momentum variation, which describe flows of viscous incompressible fluids, form a mixed elliptic–parabolic system of equations for velocity and pressure. In this case, the continuity equation includes only the velocity components. Therefore, there is no direct connection with pressure, which is performed via density in the case of compressible fluids (Wesseling 2000). The methods for solving the Navier–Stokes equations for viscous incompressible fluids are classified in pressure-based and pressure–velocity coupling methods.

To simulate flows of viscous compressible fluids in a wide range of Mach numbers in the presence of low-velocity flow subregions (stagnation and flow attachment regions, and recirculation zones), the full Navier–Stokes equations are used. The numerical solution of the Navier–Stokes equations at low Mach numbers runs into computational difficulties that result in slow convergence and reduced accuracy of the numerical solution.

To overcome the computational difficulties arising during the simulation of low-speed flows, preconditioning methods are widely used. These methods equalize the orders of the Jacobian eigenvalues in the original equations for all $M < 1$ (Choi and Merkle 1993; Volpe 1993; Weiss and Smith 1995; Turkel et al. 1997; Turkel 1987, 1993). Preconditioning modifies the terms with the time derivative in the momentum equations. When the flow reaches the steady state, the solution to the modified (preconditioned) system coincides with the solution to the original gas dynamics equations. A method that eliminates stiffness of the preconditioned Navier–Stokes equations in the computations on elongated cells in the boundary layer was proposed in Buelow et al. (1994). A review and comparative analysis of various approaches to the preconditioning of the gas dynamics equations can be found in Turkel et al. (1997), Volkov (2009), and Turkel (1993).

For the simulation of unsteady flows, the dual time-stepping method is used (Rogers and Kwak 1990; Jameson 1991). At each integration step with respect to the physical time, additional iterations on the pseudo-time are used. The integration with respect to the physical time (outer iterations) and pseudo-time (inner iterations) is performed using different schemes (e.g., implicit and explicit ones). To improve the convergence of the inner iterations, various approaches may be used (e.g., the multigrid method or residual smoothing) (Venkateswaran and Merkle 1995; Rumsey et al. 1996; Zhang et al. 2004). In Bijl et al. (2002), the dual time-stepping method is modified to enable the use of multistep difference schemes with respect to the physical time. The solution of various complex problems is described in Tang et al. (2003), and a comparison of various approaches can be found in Jameson (2009).

The analysis of stability shows that the integration step with respect to the pseudo-time is 2/3 of the physical time step (Zhao 2004), which results in the slow convergence in the case of constraints on the physical time step (e.g., when explicit difference schemes are used). To remove the constraints on the time step, implicit schemes both with respect to physical time and pseudo-time are used (fully implicit dual time-stepping schemes) (Luo et al. 2001; Zhao et al. 2002).

The preconditioning method for the Navier–Stokes equations and an approach to the computation of free convective flows of a viscous compressible fluid in the framework of the dual time-stepping procedure are developed. To construct the preconditioning matrix, the approach proposed in Weiss and Smith (1995) is used. It is based on the use of physical variables (one of these variables is the temperature).

1.3 Multigrid Method

Fluid flows are described by boundary value problems for nonlinear partial differential equations, the Euler equations for inviscid fluids, and the Navier–Stokes equations for viscous fluids. The complexity of solving the Navier–Stokes equations is due to a small parameter (Reynolds number) multiplying the higher order derivative (stiffness), on the one hand, and due to inconsistency between the right-hand part of the differential equation and the boundary conditions in the vicinity of the solid flow boundaries (boundary layer), on the other hand.

The Euler or Navier–Stokes equations discretization produces a system of difference equations with a large sparse matrix, which can be solved by various methods. If the discretization is performed on an unstructured mesh, the coefficient matrix is asymmetric and not diagonally dominant. In this case, the classical iterative methods are not effective or slowly convergent (Fedorenko 1961; Brandt 1984; Wesseling 1992). The difficulties in the numerical solution of the Euler and Navier–Stokes equations that result in slow convergence are illustrated in Fig. 1

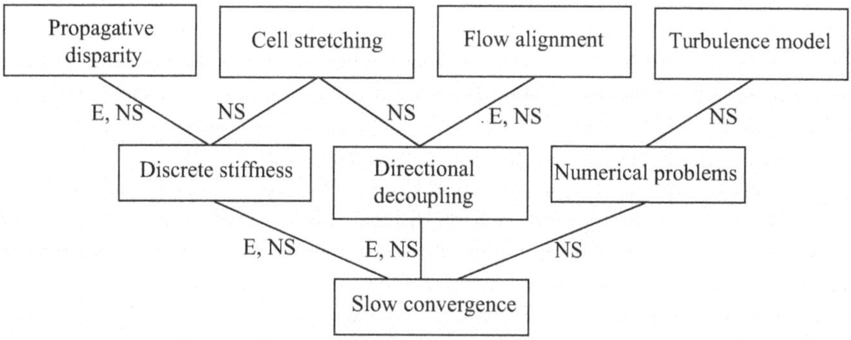

Fig. 1 Sources of slow convergence of the iterative process

from Pierce et al. (1997) (the symbols E and NS correspond to the Euler and Navier–Stokes equations, respectively). In the case of the Euler equations, the discrete stiffness is a consequence of the use of the scalar time step. The major source of difficulties in solving the Navier–Stokes equations is the use of meshes with elongated and skewed cells in the boundary layer. More difficulties arise when turbulent flows are simulated. To improve the convergence, a matrix time step and directed mesh coarsening are used (Volkov 2009, 2010).

In multigrid techniques, the conventional iterative procedure is combined with a correction of the solution on a sequence of coarse meshes (Volkov et al. 2016). One iteration step of the multigrid method (multigrid cycle) includes the following stages: pre-smoothing, calculation of the residual on the current mesh level, restriction and correction of the residual on the coarse mesh, prolongation and interpolation of the error to the fine mesh, correction of the solution on the fine mesh using the coarse mesh correction, and post-smoothing for suppressing the high-frequency error components that occur upon the interpolation to the fine mesh.

The simplest implementation of the classical multigrid techniques is the cascade method or correction scheme, which is used to solve linear systems of difference equations. It consists of an interpolation procedure (correction on a coarse mesh) with the subsequent iterative smoothing.

Nonlinear equations can be solved using the linear multigrid method in a combination with Newton's linearization. To solve the equations linearized by Newton's method, either the number of multigrid iterations is adapted for each iteration of Newton's method (in this case, a mechanism for controlling the convergence rate of Newton's method is needed) or the number of multigrid iterations at each step of Newton's method is fixed (Newton's method is simplified to make its convergence rate linear). The difficulties of this approach are due to the necessity of choosing a proper initial approximation (Mavriplis 2002).

In practice, nonlinear problems are solved using the nonlinear multigrid or the full approximation scheme. The advantages of the full approximation scheme over Newton's linearization are that there is no need for matching the internal and external iterations, and the linearization is performed within the cycle on the coarse mesh. The full approximation equation is solved on the coarse mesh, and the error and residual (which are smooth functions in distinction from the solution) are transferred to the finer mesh. Smoothing is performed in the fine mesh iterations using the Jacobi and Newton methods. The nonlinear multigrid techniques require the residual to be recalculated at each iteration step, but they avoid the calculation and storage of the Jacobian in Newton's method and the difficulties due to the choice of the initial solution in Newton's method (Brandt 1977).

The implementation of the multigrid method is reduced to specifying the number of smoothing iteration steps for each mesh level, the number of recursive calls of the method on each mesh level, the number of iteration steps on the coarsest mesh, and the number of the mesh level at which the multigrid cycle starts. At $\gamma = 1$, one iteration of the method gives a V-cycle, and at $\gamma = 2$, it gives a W-cycle. The difference equations on the coarsest mesh level are solved using direct or iterative solution methods. In the general case, the number of smoothing iterations and the number of recursive calls of the multigrid method may depend on the mesh level (F-cycle).

The efficiency of multigrid techniques increases if they are used in the form of the full multigrid method (FMG). Smoothing on the fine mesh and correction on the coarse mesh levels are accompanied by nested iterations. The full multigrid method is more efficient than the simple multigrid method due to the combined use of the solution obtained on the coarse mesh levels as an initial solution for solving the problem on the fine mesh. The full multigrid method is used for solving both linear and nonlinear problems (based on the full approximation scheme).

Modern numerical methods use multilevel multigrid techniques with the explicit (geometric approach, GMG) and implicit (algebraic approach, AMG) construction of the mesh sequence (Brandt 1977; Trottenberg et al. 2001). By using several discretization levels, the multigrid method makes it possible to resolve conflicts between high-frequency (fine mesh discretization) and low-frequency (coarse mesh discretization) components of the solution. They achieve high efficiency by reducing the amount of computations needed to obtain a numerical solution (Trottenberg et al. 2001; Wesseling and Oosterlee 2001).

Commercial CFD software includes both implementations of geometric and algebraic multigrid techniques. To construct the sequence of meshes, the geometric approach uses the agglomeration method. However, its implementation ignores the problem nature, which results in difficulties related to the convergence of the numerical solution in the boundary layer when meshes with elongated and skewed cells are used. The algebraic approach uses the piecewise constant interpolation within the Galerkin approach to construct discrete operators on the coarse mesh levels.

The multigrid method only establishes the structure of the computational algorithm, but its efficiency depends on the adaptation of its components to the problem in question. The geometric and algebraic multigrid techniques implemented within the finite volume approach to the discretization of the Euler and Navier–Stokes equations on unstructured meshes are compared.

2 Governing Equations

The calculations are based on the compressible flow CFD code designed with air as the working fluid. The perfect gas law is used to link the density, pressure, and temperature.

2.1 Cartesian Coordinates

In Cartesian coordinates (x, y, z), the unsteady viscous compressible flow is described by the equation

$$\frac{\partial Q}{\partial t} + \frac{\partial F_x}{\partial x} + \frac{\partial F_y}{\partial y} + \frac{\partial F_z}{\partial z} = 0. \tag{1}$$

Equation (1) is supplemented with the ideal gas equation of state

$$p = (\gamma - 1)\rho\left[e - \frac{1}{2}\left(v_x^2 + v_y^2 + v_z^2\right)\right].$$

The vector of conservative variables Q and the flux vectors F_x, F_y, and F_z have the form

$$Q = \begin{pmatrix} \rho \\ \rho v_x \\ \rho v_y \\ \rho v_z \\ \rho e \end{pmatrix},$$

$$F_x = \begin{pmatrix} \rho v_x \\ \rho v_x v_x + p - \tau_{xx} \\ \rho v_x v_y - \tau_{xy} \\ \rho v_x v_z - \tau_{xz} \\ (\rho e + p) v_x - v_x \tau_{xx} - v_y \tau_{xy} - v_z \tau_{xz} + q_x \end{pmatrix},$$

$$F_y = \begin{pmatrix} \rho v_y \\ \rho v_y v_x - \tau_{yx} \\ \rho v_y v_y + p - \tau_{yy} \\ \rho v_y v_z - \tau_{yz} \\ (\rho e + p) v_y - v_x \tau_{yx} - v_y \tau_{yy} - v_z \tau_{yz} + q_y \end{pmatrix},$$

$$F_z = \begin{pmatrix} \rho v_z \\ \rho v_z v_x - \tau_{zx} \\ \rho v_z v_y - \tau_{zy} \\ \rho v_z v_z + p - \tau_{zz} \\ (\rho e + p) v_z - v_x \tau_{zx} - v_y \tau_{zy} - v_z \tau_{zz} + q_z \end{pmatrix}.$$

The components of the viscous stress tensor and the heat flux components are given by

$$\tau_{ij} = \mu\left(\frac{\partial v_i}{\partial x_j} + \frac{\partial v_j}{\partial x_i} - \frac{2}{3}\frac{\partial v_k}{\partial x_k}\delta_{ij}\right), \quad q_i = -\lambda\frac{\partial T}{\partial x_i}.$$

Here, t is time; ρ is the density; v_x, v_y, and v_z are the velocity components in the x, y, and z directions; p is the pressure; e is the total energy per unit mass; T is the temperature; and γ is the ratio of specific heat capacities.

The molecular viscosity, μ, is a function of temperature. It is modeled with Sutherland's law

$$\frac{\mu}{\mu_*} = \left(\frac{T}{T_*}\right)^{3/2} \frac{T_* + S_0}{T + S_0},$$

where μ_* and T_* are a reference viscosity and temperature, and S_0 is a constant determined experimentally, so that $\mu_* = 1.7894 \times 10^{-5}$ kg/(m s), $T_* = 273.11$ K, and $S_0 = 110.56$ K for air.

The thermal conductivity, λ, is linked to the specific heat capacity at constant pressure, c_p, and the Prandtl number, Pr, so that $\lambda = c_p \mu / Pr$, and $Pr = 0.7$ for air.

2.2 Boundary Conditions

To obtain a well-posed problem, appropriate boundary conditions must be imposed on the boundary of computational domain.

Boundary conditions used in calculations include inviscid wall (slip and no-penetration boundary conditions), viscous wall (no-slip and no-penetration boundary conditions), free stream, inflow, outflow, periodic, and centreline boundary conditions.

No parameters are required for inviscid wall (the appropriate solid wall boundary condition is zero velocity normal to the wall). For the viscous wall, three velocity components (two in 2D case) or the rotation speed for an annular model can be specified if the wall is moving. A constant wall temperature may be applied or the wall can be adiabatic.

For a free stream boundary, the density, pressure, Mach number, and angle of attack and yaw angle must be specified.

To define the inflow boundary conditions, for 2D calculations, one needs to provide a total pressure, a total temperature, and an inflow angle. In 3D cases, 1D profiles of these parameters can be specified, and both the tangential and radial flow angles are required.

For the outflow boundary, a static pressure and reverse flow temperature are required. If the reverse flow temperature is set at zero (advised if unknown), the reverse flow temperature would be taken identical to the temperature of the flow going out of the boundary.

Periodic boundary conditions are specified in circumferential directions, and symmetric boundary conditions are specified on the centreline.

3 Low Mach Preconditioning

Preconditioning based on the use of physical variables is used to speed up the convergence of time marching to a steady state and to improve the accuracy of the steady-state solution.

3.1 Low Mach Flows

Consider the linearized Euler equations

$$\frac{\partial U}{\partial t} + A\frac{\partial U}{\partial x} = 0,$$

where $A = \partial F/\partial U$ is the Jacobian. The eigenvalues of the Jacobian are $\lambda_1 = u$, $\lambda_2 = u - c$, and $\lambda_3 = u + c$. The condition number of the matrix A has the form

$$k(A) = \rho(A)\,\rho\left(A^{-1}\right),$$

where ρ is the spectral radius. For $k \gg 1$, the matrix A is ill-conditioned, which causes difficulties in solving the corresponding system of difference equations.

Assuming for the sake of simplicity that $u > 0$, the condition number is represented as

$$k(A) = (u+c)\max\left\{\frac{1}{u}, \frac{1}{|u-c|}\right\} = \max\left\{1 + \frac{1}{M}, \frac{M+1}{|M-1|}\right\}.$$

The Jacobian is ill-conditioned at the sonic point (as $M \to 1$) and in low Mach number flows (as $M \to 0$).

The computational difficulties near the sonic point can be overcome by adding dissipative terms (Harten 1983).

For the explicit Euler scheme, the stability condition has the form $\Delta t \leqslant \Delta x/(u+c)$. In subsonic flows, the physical time step is of the order of the characteristic timescale $\Delta\tau = \Delta x/u$. Comparing the numerical and physical time steps gives

$$\frac{\Delta t}{\Delta \tau} \leqslant \frac{M}{1+M}.$$

At low Mach numbers, the numerical time step is less than the physical one and the governing equations are stiff.

3.2 Finite Volume Method

In conservative variables, the equation describing an unsteady viscous compressible flow is written as

$$\frac{\partial U}{\partial t} + \nabla \cdot [F(U) + G(U, \nabla U)] = 0, \tag{2}$$

where U is the vector of conservative variables, $F(U)$ is the vector of inviscid fluxes, and $G(U, \nabla U)$ is the vector of viscous fluxes. For simplicity, the source term in Eq. (2) is omitted.

Integrating Eq. (2) over the control volume V with boundary ∂V, whose orientation is specified by the outer unit normal n, and applying the Gauss theorem yield

$$\frac{\partial}{\partial t} \int_V U \, d\Omega + \int_{\partial V} F(U) \cdot n \, dS + \int_{\partial V} G(U) \cdot n \, dS = 0. \tag{3}$$

The Navier–Stokes equations written in the form of (3) are spatially discretized using the vertex-centered finite volume method on an unstructured mesh. The explicit Euler method or the explicit two-step Runge–Kutta method is used for time integration. Inviscid fluxes are discretized by applying the Roe scheme, while viscous fluxes are discretized with the help of a second-order accurate centered scheme. Some details of the implementation of the finite volume method are discussed in Volkov (2009, 2011).

3.3 Preconditioning Matrix

At low Mach numbers, the flow is described by an equation written in physical variables. The new set of physical variables is defined as

$$Q = \begin{pmatrix} p \\ v_x \\ v_y \\ v_z \\ T \end{pmatrix}.$$

For incompressible flows, the velocity and temperature gradients are required for computing viscous flows, and the pressure gradient is needed for the solution interpretation. In the computations, the pressure excess $p - p_*$ is used, which is taken into account when the density is found from the equation of state (in the basic equations, the pressure is under the integral sign).

The equations describing the flow at low Mach numbers are written in physical variables

$$\Gamma \frac{\partial}{\partial t} \int_{V_i} Q \, dV + \int_{\partial V_i} F \, dS = 0. \quad (4)$$

The preconditioning matrix is represented as

$$\Gamma = \begin{pmatrix} \Theta & 0 & 0 & 0 & \rho_T \\ \Theta v_x & \rho & 0 & 0 & \rho_T v_x \\ \Theta v_y & 0 & \rho & 0 & \rho_T v_y \\ \Theta v_z & 0 & 0 & \rho & \rho_T v_z \\ \Theta H - 1 & \rho v_x & \rho v_y & \rho v_z & \rho_T H + \rho c_p \end{pmatrix},$$

where $H = c_p T + |v|^2/2$, and c_p is the specific heat capacity at constant pressure. The density derivatives with respect to pressure at constant temperature and with respect to temperature at constant pressure are determined by the relations

$$\rho_p = \left.\frac{\partial \rho}{\partial p}\right|_T, \quad \rho_T = \left.\frac{\partial \rho}{\partial T}\right|_p.$$

The parameter Θ is given by

$$\Theta = \left(\frac{1}{U_r^2} - \frac{\rho_T}{\rho c_p} \right),$$

where U_r is the propagation velocity of pressure perturbations.

For an ideal gas, $\rho_p = 1/(RT) = \gamma/c^2$, where c is the speed of sound, while, for constant density flows, $\rho_p = 0$, which leads to pressure perturbations propagating at an infinite speed in an incompressible fluid. To eliminate the singularity, ρ_p is assumed to be inversely proportional to the local velocity squared (the propagation speed of pressure perturbations is equal to the local velocity). As a result, the eigenvalues of the Jacobian become of the same order, and the Euler and Navier–Stokes equations at low Mach numbers are no longer stiff.

For an ideal gas, the parameter U_r is defined as

$$U_r = \begin{cases} \varepsilon c, & \text{if} \quad |v| < \varepsilon c, \\ |v|, & \text{if} \quad \varepsilon c < |v| < c, \\ c, & \text{if} \quad |v| > c. \end{cases}$$

The preconditioning mechanism is used when the local velocity of the flow is less than the speed of sound (then $U_r = u$). In a supersonic flow region, preconditioning is not used $(U_r = c)$ and preconditioned system (4) passes into system (3) for compressible gas flows. For $u \ll c$ (for example, at stagnation points), U_r is bounded by a certain value.

For incompressible and variable density fluids (for example, when the Boussinesq approximation is used for free convective flows), the parameter U_r is specified as

$$U_r = \begin{cases} \varepsilon U_{max}, & \text{if} \quad |v| < \varepsilon U_{max}, \\ |v|, & \text{if} \quad |v| > \varepsilon U_{max}, \end{cases}$$

where U_{max} is the maximum flow velocity.

To represent viscous fluxes correctly, it is necessary that the propagation speed of acoustic perturbations associated with the eigenvalues of the modified system of equations be no less than the propagation speed of viscous perturbations (Weiss and Smith 1995). A switch is introduced in control volumes where the velocity is so low that the eigenvalues of the preconditioned system become less than the propagation speed of viscous perturbations (for example, in boundary and mixing layers, where viscous effects are dominant). The parameter U_r is replaced by

$$U_r^* = \max\left\{U_r, \frac{\mu}{\rho \Delta x}\right\}.$$

The eigenvalues of the Jacobian of the preconditioned system are found by solving the characteristic equation $|A - \lambda \Gamma| = 0$, where $A = \partial F_x / \partial Q$. Hence

$$\lambda_{1,2,3} = v_x, \quad \lambda_{4,5} = v_x' \mp c',$$

where

$$v_x' = v_x(1 - \alpha), \quad c' = \left(\alpha^2 v_x^2 + U_r^2\right)^{1/2}.$$

The parameter α is given by the relation

$$\alpha = \frac{1}{2}\left(1 - \beta U_r^2\right), \quad \beta = \rho_p + \frac{\rho_T}{\rho c_p}.$$

In supersonic flows, $U_r = c$ and $\alpha = 0$, and the eigenvalues are of the same order and equal to the eigenvalues of the original system ($\lambda_{1,2,3} = v_x$, $\lambda_{4,5} = v_x \mp c$). At low Mach numbers ($|v_x| \ll c$), $\alpha \to 1/2$ as $U_r \to 0$, and the eigenvalues have the same order and tend to the value

$$v_x' \mp c' = \frac{1}{2} v_x (1 \mp \sqrt{5}).$$

In the case of incompressible flows, $\beta = 0$ and $\alpha = 1/2$ (there is no dependence on U_r), the eigenvalues are also of the same order, and the stiffness of the system is eliminated as $M \to 0$.

3.4 Discretization of Preconditioning Equations

A finite volume discretization of Eq. (4) yields

$$\Gamma \frac{dQ_i}{dt} + \frac{1}{V_i} \sum_{j}^{N_i} (F_{ij} + G_{ij}) S_{ij} = 0. \tag{5}$$

The inviscid fluxes on a control volume face are determined using the relation

$$F_{i+1/2} = \frac{1}{2}(F_R + F_L) - \frac{1}{2}|A|_{i+1/2} \Delta U.$$

The viscous fluxes G_{ij} are computed in the same way as without preconditioning (Volkov 2009). The second order in space is achieved by interpolating the gradients to a control volume edge.

The shortcomings of the numerical algorithms as $M \to 0$ are eliminated by preconditioning the dissipative term of the difference scheme (Turkel et al. 1993). The dissipative term is represented as

$$|A|\Delta U = \Gamma \Gamma^{-1} \frac{\partial F_x}{\partial U} \Delta U = \Gamma \left(\Gamma^{-1} \frac{\partial F_x}{\partial Q} \right) \Delta Q = \Gamma |A_\Gamma| \Delta Q,$$

where $A_\Gamma = \Gamma^{-1}(\partial F_x / \partial Q)$. The eigenvalues λ_4 and λ_5 are corrected near the sonic point (Harten 1983). In contrast to the Roe scheme, the arithmetic means of parameter values are used to compute the eigenvalues and coefficients of the matrices A_Γ and Γ.

Diagonalizing the Jacobian matrix $A_\Gamma = M_\Gamma \Lambda_\Gamma M_\Gamma^{-1}$ of the preconditioned system, the difference scheme is written in the form

$$F_{i+1/2} = \frac{1}{2}(F_R + F_L) - \frac{1}{2} \Gamma M_\Gamma |\Lambda_\Gamma| M_\Gamma^{-1} \Delta Q.$$

For explicit difference schemes, the dissipative term simplifies to

$$|A_\Gamma|\Delta Q = M_\Gamma|\Lambda_\Gamma|M_\Gamma^{-1}\Delta Q$$
$$= \begin{pmatrix} c^*\Delta p + M^*(\rho U_r^2\Delta v_x - \alpha v_x \Delta p) \\ c^*\Delta v_x + M^*(\alpha v_x \Delta v_x + \Delta p/\rho) \\ |\lambda_1|\Delta v_y \\ |\lambda_1|\Delta v_z \\ \left[c^*\Delta p + M^*(\rho U_r^2\Delta v_x - \alpha v_x\Delta p) + |\lambda_1|(\rho c_p \Delta T - \Delta p)\right]/(\rho c_p) \end{pmatrix},$$

where

$$M^* = \frac{1}{2c'}(|\lambda_5| - |\lambda_4|), \quad c^* = \frac{1}{2}(|\lambda_5| + |\lambda_4|).$$

Here, Δp, Δv_x, Δv_y, Δv_z, and ΔT denote the jumps in pressure, velocities, and temperature on the control volume edge, respectively. The eigenvalues of the Jacobian are given by $\lambda_1 = v_x$, $\lambda_4 = u' - c'$, and $\lambda_5 = u' + c'$.

The order of accuracy in space is increased by applying an approach based on linear distributions of the parameters over a control volume. The variables at the center of the control volume Q_i are interpolated to an edge

$$Q'_\mathrm{L} = Q_\mathrm{L} + (\nabla Q)_\mathrm{L}|\Delta r_\mathrm{L}|, \quad Q'_\mathrm{R} = Q_\mathrm{R} + (\nabla Q)_\mathrm{R}|\Delta r_\mathrm{R}|,$$

where ∇Q is the gradient at the control volume center, and Δr_L and Δr_R are the vectors directed from the centers of neighboring control volumes L and R to the midpoint of the separating edge.

3.5 Time-Marching Scheme

Equation (5) is written in semi-discrete form

$$\frac{\mathrm{d}Q_i}{\mathrm{d}t} = -\Gamma^{-1}R. \tag{6}$$

Discretizing the residual yields

$$R = \frac{1}{V_i}\sum_j^{N_i} F_{ij}S_{ij}.$$

The time differencing of Eq. (6) is based on the explicit Euler method. Specifically, at the time level $n+1$, the variables are determined using the relations

$$Q_i^{n+1} = Q_i^n + \Delta t \Gamma^{-1}(Q_i^n) R(Q^n).$$

At use of the explicit two-step Runge–Kutta method for a discretization of Eq. (6) in time variables on a layer $n+1$ in time are from relations

$$Q_i^* = Q_i^n + \Delta t \Gamma^{-1}(Q_i^n) R(Q^n);$$
$$Q_i^{n+1} = Q_i^n + \Delta t \Gamma^{-1}(Q_i^*) R(Q^*).$$

When steady-state problems are solved by applying time-marching scheme, the time derivative in Eq. (6) vanishes as $t \to \infty$. Accordingly, the solution of Eq. (6) is regarded as the solution to the original system of equations. The time step is chosen as based on the highest propagation velocity of pseudo-pressure perturbations or viscous perturbations and is calculated using the formula

$$\Delta t = \min\left\{ \mathrm{CFL} \frac{\Delta x}{|u'| + c'}, \sigma \frac{\rho \Delta x^2}{\mu} \right\},$$

where CFL is the Courant number, σ is a parameter bounding the time step in the case of viscous problems, and Δx is the shortest distance from the control volume center to the face.

3.6 Numerical Examples

A series of CFD simulations in a wide range of Mach numbers were used to analyze the convergence rate and accuracy of steady-state solutions of the original and preconditioned gas dynamics equations. These equations were integrated until a steady-state solution was reached. As a convergence criterion, the residual norm is used

$$L_2(U) = \left[\sum_{i=1}^{N} \left(U_i^{n+1} - U_i^n \right)^2 \right]^{1/2}.$$

The numerical solution was said to converge if $L_2 \leqslant \varepsilon$, where ε is a small prescribed quantity ($\varepsilon \sim 10^{-16}$ was used in the computations).

3.6.1 Nozzle Flow

Consider the inviscid compressible gas flow in a channel with a variable cross-sectional area. The nozzle profile is described by the relation

$$y = \left(\frac{1+x^2}{\pi}\right)^{1/2},$$

where $x \in [-0.3, 1]$.

The flow regime in the nozzle is determined by the relation between the total pressure p_1 and the outlet pressure p_2 ($\Delta p = p_1 - p_2$ was specified in the computations). In version 1 (the pressure drop is less than a critical value), the flow has a shock wave sitting at the outlet. In version 2 (the pressure drop is greater than the critical value), the gas accelerates from subsonic inlet conditions to a speed in the critical cross section that depends on the specified pressure drop and then decelerates. The computations were performed on a mesh of 100 cells at CFL $= 0.95$.

In version 1, the total pressure ($p_1 = 10^6$ Pa) and the total temperature ($T_1 = 300$ K) in the inlet section are specified, while the static pressure ($p_2 = 8 \times 10^5$ Pa) is fixed on the outlet section. The flow at the inlet of the computational domain is subsonic. In the converging section of the nozzle, the gas accelerates, reaches the speed of sound in the critical cross section, and moves further at a supersonic speed. In the expanding section of the nozzle, there develops a normal shock wave, behind which the flow becomes subsonic. With the given parameters, $M > 0.4$ in the entire computational domain, so the solutions of the original and preconditioned equations coincide.

In version 2, the gas accelerates in the subsonic section and decelerates in the supersonic one. Figures 1, 2, 3, 4, and 5 show the convergence rates of the time-marching procedure in the form of the residual norm as a function of the number of time steps for the original and preconditioned equations. Lines 1 and 2 depict the residuals caused by discretizing the momentum and pressure equations, respectively.

At $\Delta p = 1600$ Pa, the Mach number in the critical cross section is approximately $M = 0.3$. The convergence of the numerical solution is shown in Fig. 2. The flow

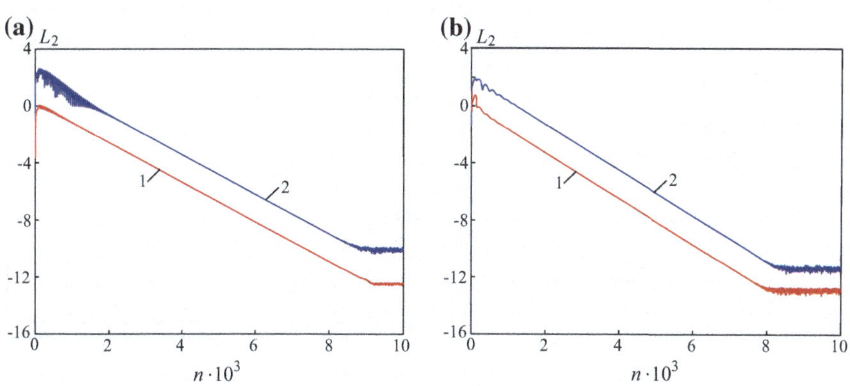

Fig. 2 Convergence rate of the solutions of the original (**a**) and preconditioned (**b**) equations at $\Delta p = 1600$ Pa

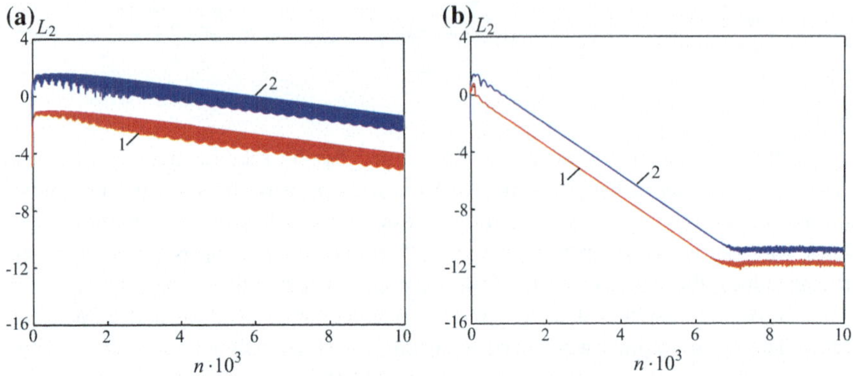

Fig. 3 Convergence rate of the solutions of the original (**a**) and preconditioned (**b**) equations at $\Delta p = 175$ Pa

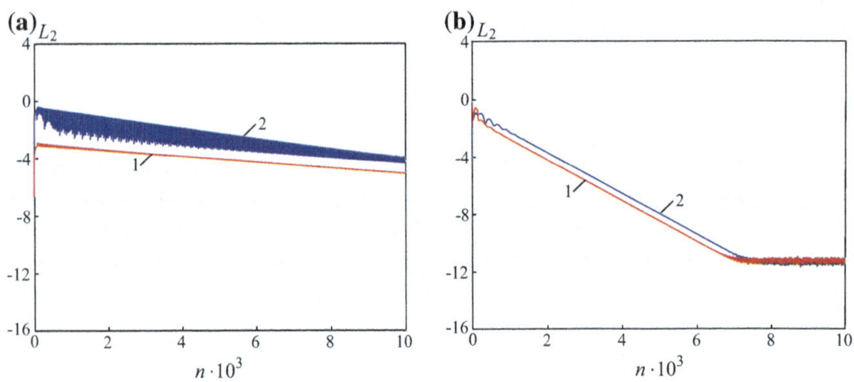

Fig. 4 Convergence rate of the solutions of the original (**a**) and preconditioned (**b**) equations at $\Delta p = 1.8$ Pa

Fig. 5 Convergence rate of the solutions of the original (**a**) and preconditioned (**b**) equations at $\Delta p = 0.02$ Pa

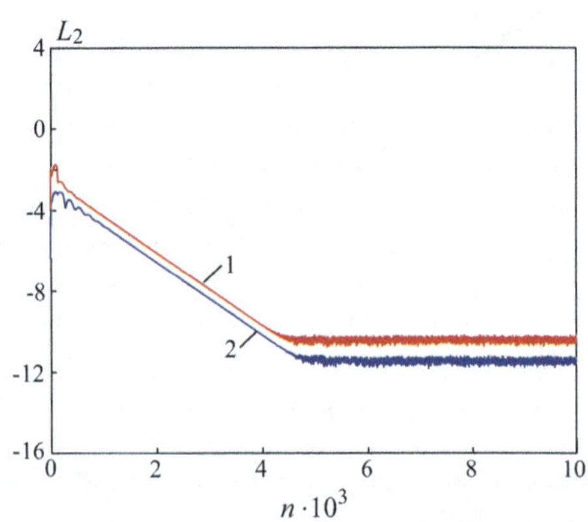

velocity remains sufficiently high in the entire computational domain, so the solutions based on two numerical procedures nearly coincide. The residual patterns are also similar in both cases. However, the solution of the original equations exhibits a lower residual (Fig. 2a) than that of the preconditioned equations (Fig. 2b). The solution produced by time marching reaches a steady state after about 9×10^3 time steps in the case of the original equations and after 8×10^3 time steps in the case of preconditioning.

When the pressure drop is reduced to $\Delta p = 175$ Pa, the Mach number in the critical cross section becomes $M = 0.1$. The convergence of the numerical solution is shown in Fig. 3. The solutions of the original and preconditioned Euler equations are now different. Due to the developed preconditioning approach, the prescribed residual level is achieved after about 7×10^3 time steps (Fig. 3b). For the original equations, the convergence rate of the time-marching procedure is rather slow, and the prescribed residual level is not achieved after 10^4 time steps (Fig. 3a).

As the pressure drops further to $\Delta p = 1.8$ Pa, the Mach number in the critical cross section becomes $M = 0.01$. The convergence of the numerical solution is shown in Fig. 4. The Mach number distributions along the nozzle axis differ substantially in the case of the original and preconditioned equations (the distributions of flow characteristics are not shown). Due to the preconditioning procedure, the convergence pattern becomes nearly independent of the pressure drop (Fig. 4b), while the convergence of the solution to the original equations degrades (Fig. 4a).

When the pressure drop is reduced to $\Delta p = 0.02$ Pa, the Mach number in the critical cross section is $M = 0.001$. The solution of the original equations becomes divergent. The convergence of the solution of the preconditioned equations is illustrated in Fig. 5. The prescribed residual is achieved after 4.8×10^3 time steps.

The variation in the residual shown in Fig. 5 is similar to the convergence patterns presented in Figs. 3 and 4, which suggests that the time-marching procedure depends weakly on the Mach number in the computational domain. Relying on the developed approach, the characteristics of the nozzle flow from essentially subsonic to supersonic velocities are computed.

3.6.2 Flow Through a Channel with a Bump

Consider the flow through a plane channel with a bump. The length to height ratio in the channel is $L/H = 4$, and the maximum height of the bump (which is a circular arc) is $0.1H$ (the maximum bump is 10% of the channel width). The computations were performed on a mesh of 120×20 cells (Fig. 6) with 60 nodes placed on the bump surface.

Flows through a channel with a bump were computed, for example, in Choi and Merkle (1985), Eidelman et al. (1984). Specifically, the implicit Euler time differencing and the Beam–Warming scheme for discretizing inviscid fluxes were used in Choi and Merkle (1985). The computations in Eidelman et al. (1984) were based on Godunov method and were performed in a wide range of Mach numbers.

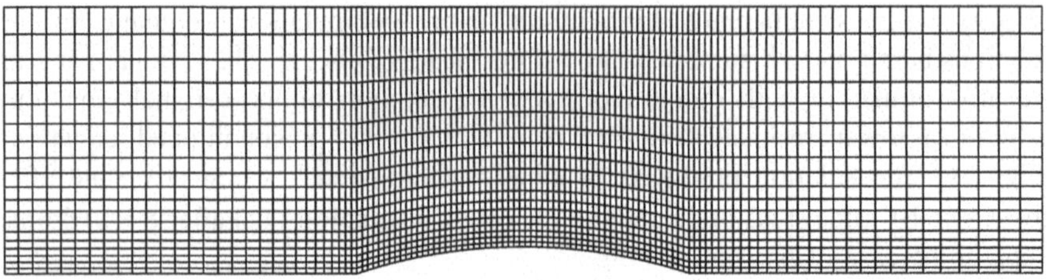

Fig. 6 Mesh

The velocity ($U = 3.47$ m/s), pressure ($p = 10^5$ Pa), and temperature ($T = 300$ K) were set in the inlet section of the channel, while free outflow boundary conditions were specified in the outlet section. The inlet boundary conditions corresponded to $M = 0.01$. A steady-state solution of the problem was obtained by taking 5000 time steps of the time-marching procedure.

Figure 7 displays contours of the velocity magnitude. In contrast to the solution of the original equations, a velocity distribution symmetric to the vertical axis is obtained in the case of preconditioning.

The convergence rate of the time-marching procedure is shown in Fig. 8. The original equations were solved in conservative variables, while the preconditioned equations were computed in physical variables. Lines 1 and 2 depict the residuals (in physical variables) caused by discretizing the momentum equation, while line 3 shows the residual caused by discretizing the pressure equation. In the case of preconditioning, the prescribed residual is obtained after about 3500 iteration steps. For the original equations, the residuals with respect to velocity and pressure are

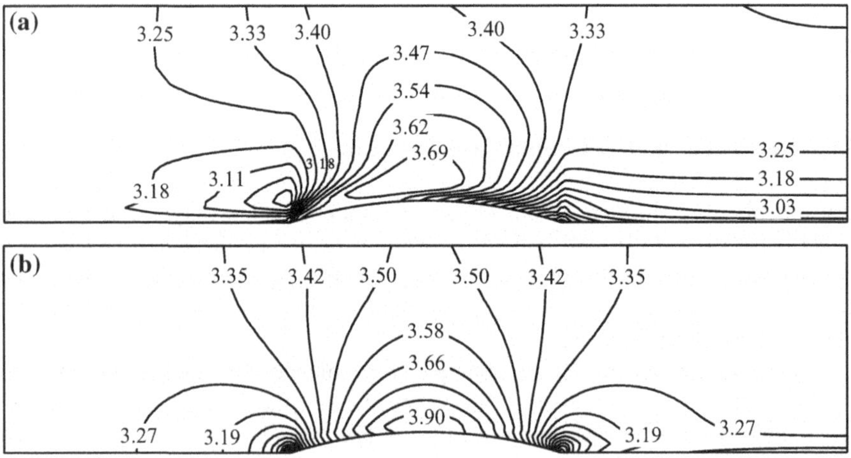

Fig. 7 Contours of the velocity magnitude in the case of the original (**a**) and preconditioned (**b**) equations

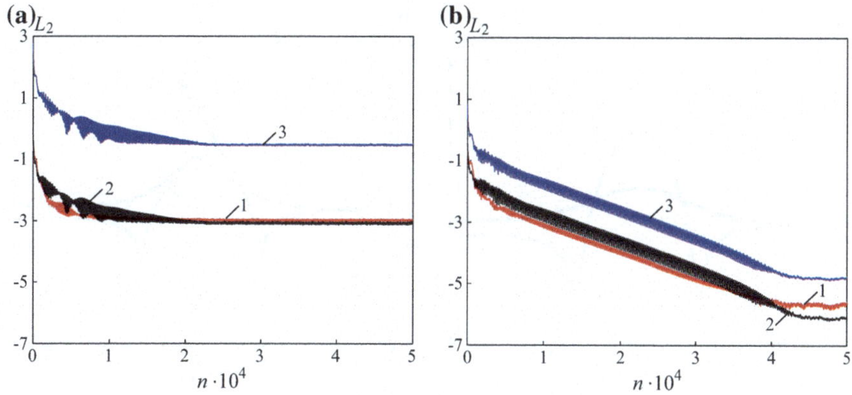

Fig. 8 Convergence rate of the solutions of the original (**a**) and preconditioned (**b**) equations

two orders of higher magnitude, and the convergence rate nearly ceases to vary after 4000 time steps.

To test the performance and accuracy of the numerical method on a wide range of Mach numbers, the computations were performed in subsonic, transonic, and supersonic regimes. More specifically, the flow was computed in a channel with a 10% bump (as in the underlying version) on a mesh of 144×32 cells at $M = 0.5$ (subsonic) and $M = 0.675$ (transonic) and in a channel with a 4% bump on a mesh of 220×60 cells at $M = 1.65$ (supersonic).

For subsonic and supersonic regimes, Fig. 9 shows contours of the velocity magnitude at various inlet Mach numbers. For relatively low inlet Mach numbers, the flow is nearly symmetric to the vertical axis (Fig. 9a). The weak asymmetry of

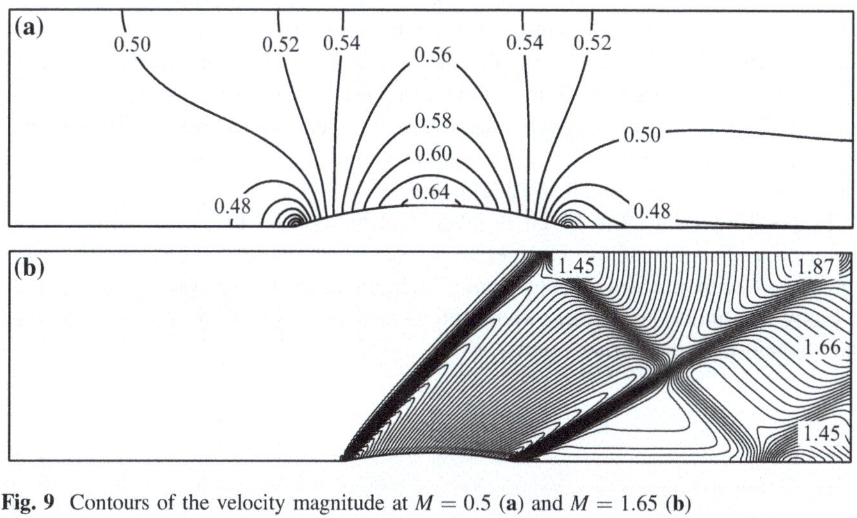

Fig. 9 Contours of the velocity magnitude at $M = 0.5$ (**a**) and $M = 1.65$ (**b**)

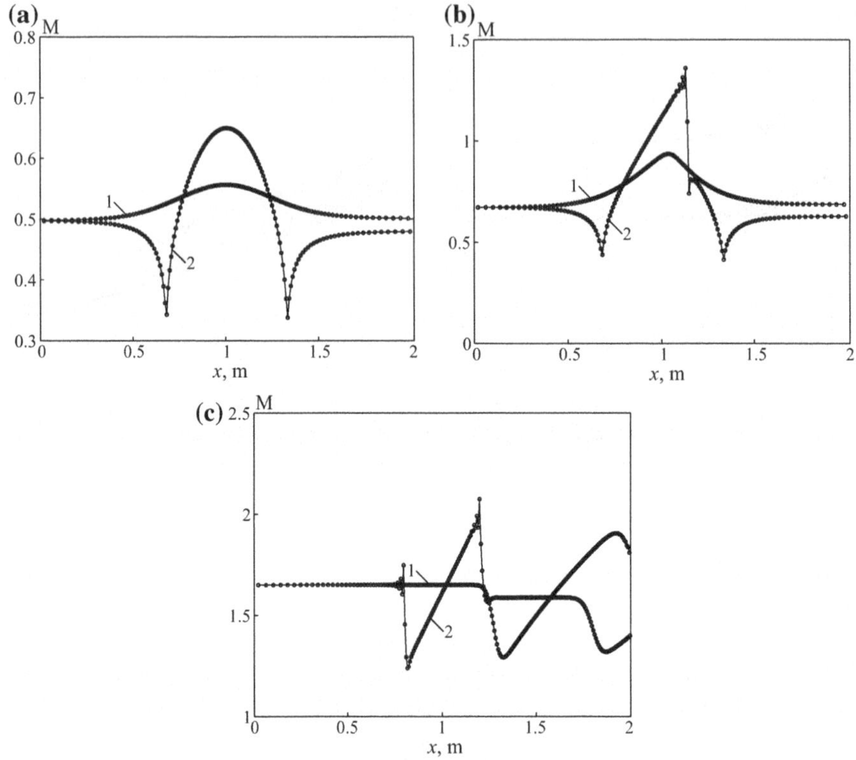

Fig. 10 Mach number distribution along the upper (line 1) and lower (line 2) channel walls at $M = 0.5$ (**a**), $M = 0.765$ (**b**), and $M = 0.1.65$ (**c**)

the flow is associated with the leading and trailing edges of the bump (a horseshoe vortex of weak intensity develops behind the bump). To eliminate these shortcomings, the flow characteristics near the corner points are computed by interpolating the flow parameters from the interior nodes of the computational domain (Eidelman et al. 1984). At high inlet Mach numbers, shock waves develop and interact in the flow (Fig. 9b). The shock angle and the level lines agree well with the numerical data presented in Eidelman et al. (1984).

Figure 10 presents the Mach number distributions on the upper (line 1) and lower (line 2) walls of the channel in various flow regimes. These distributions agree well with numerical data from Eidelman et al. (1984) (as in the case of velocity magnitude level lines, weak differences are observed on the lower wall of the channel near the corner points).

4 Computation of Free Convective Flows

The specific features of the free convective flows simulation of viscous compressible fluids based on the full Navier–Stokes equations are considered.

4.1 *Finite Volume Method*

In terms of conservative variables, the equation describing the unsteady flow of viscous compressible fluid is

$$\frac{\partial U}{\partial t} + \nabla \cdot F = H, \tag{7}$$

where U, F, and H form the vector of conservative variables, the vector of fluxes, and the source term, respectively. The source term H takes into account mass forces. It has the form

$$H = \left(0, \rho g_x, \rho g_y, \rho g_z, 0\right)',$$

where $g = \{g_x, g_y, g_z\}$ is the acceleration due to gravity.

By integrating Eq. (7) with respect to the control volume V with the boundary ∂V whose orientation is specified by the outward unit normal n and using the Gauss theorem yield

$$\frac{\partial}{\partial t}\int\limits_V U\,\mathrm{d}V + \int\limits_{\partial V} F \cdot n\,\mathrm{d}S = \int\limits_V H\,\mathrm{d}V. \tag{8}$$

The Navier–Stokes equations of form (8) are discretized using the finite volume method on an unstructured mesh (the control volume is centered with respect to the mesh node). For time discretization, explicit and implicit difference schemes are used. The inviscid fluxes are discretized using Roe scheme, and the viscous fluxes are discretized using a centered scheme of the second-order accuracy. Details of the finite volume method are discussed in Volkov (2009).

4.2 *Preconditioning Method*

For the simulation of low-speed inviscid flows, the physical variables, Q, rather than the conservative variable, U, are used. The choice of the primitive variables ensures the accurate computation of velocity gradients as viscous fluxes are discretized and of the pressure gradient as convective fluxes are discretized.

The equations describing the flow of viscous compressible fluid at low Mach numbers are written in terms of the physical variables as

$$\Gamma \frac{\partial}{\partial t} \int\limits_{V} Q \, dV + \int\limits_{\partial V} F \, dS = 0. \tag{9}$$

The preconditioning matrix has the form

$$\Gamma = \begin{pmatrix} \Theta & 0 & 0 & 0 & \rho_T \\ \Theta v_x & \rho & 0 & 0 & \rho_T v_x \\ \Theta v_y & 0 & \rho & 0 & \rho_T v_y \\ \Theta v_z & 0 & 0 & \rho & \rho_T v_z \\ \Theta H - \delta & \rho v_x & \rho v_y & \rho v_z & \rho_T H + \rho c_p \end{pmatrix},$$

where $H = c_p T + |v|^2/2$, c_p is the specific heat capacity at constant pressure, and $\delta = 0$ for the perfect gas and $\delta = 1$ for the inviscid fluid. The derivative of density with respect to pressure at constant temperature and the derivative of density with respect to temperature at constant pressure are found by

$$\rho_p = \left.\frac{\partial \rho}{\partial p}\right|_T, \quad \rho_T = \left.\frac{\partial \rho}{\partial T}\right|_p.$$

Parameter Θ is given by the formula

$$\Theta = \left(\frac{1}{U_r^2} - \frac{\rho_T}{\rho c_p} \right),$$

where U_r is the velocity of pressure perturbation propagation. The parameter U_r is chosen so as to make the eigenvalues of the Jacobian the same order of magnitude as the convective and diffusive scales (Weiss and Smith 1995). Hence, $\rho_p = 1/\mathrm{RT} = \gamma/c^2$ for the perfect gas, and $\rho_p = 0$ for flows with a constant density.

4.3 Dual Time-Stepping Scheme

In the unsteady flows of viscous compressible fluids computations, the time derivative in Eq. (9) does not tend to zero as $t \to \infty$. For this reason, a term with the derivative with respect to the pseudo-time is added to Eq. (8) (the dual time stepping scheme). Equation (9) takes the form

$$\frac{\partial}{\partial t}\int\limits_V U\,\mathrm{d}V + \Gamma\frac{\partial}{\partial \tau}\int\limits_V Q\,\mathrm{d}V + \int\limits_{\partial V} F\,\mathrm{d}S = \int\limits_V H\,\mathrm{d}V, \tag{10}$$

where t is the physical time and τ is the pseudo-time. The solution to Eq. (10) is a solution to the original system of Eq. (8) if the term with the derivative with respect to the pseudo-time tends to zero at each physical time step. To execute one physical time step, several pseudo-time steps are executed until the solution stops to change (until a given residual level is reached).

4.3.1 Explicit Scheme

Discretization of the physical time derivatives in Eq. (10) with the first or second order of accuracy and the derivatives with respect to pseudo-time with the first order of accuracy, for the control volume i, yield (to simplify the notation, the index i is missed)

$$\frac{a_0 U^k - a_1 U^n + a_2 U^{n-1}}{\Delta t} + \Gamma^n \frac{Q^k - Q^{k-1}}{\Delta \tau} = R(Q^{k-1}). \tag{11}$$

Here, $a_0 = a_1 = 1/2$ and $a_2 = 0$ for the first-order scheme, and $a_0 = 3/2$, $a_1 = 2$, and $a_2 = 1/2$ for the second-order scheme. The residual is found by the formula

$$R = -\frac{1}{V}\left[\sum_{j=1}^{N_i} F_j S_j - H\right].$$

Here, Γ^n is the preconditioning matrix calculated in the control volume i on the time layer n. The index n refers to the physical time layer, and k is the internal index of pseudo-time iterations.

Rearranging the terms in Eq. (11) gives

$$\left(\Gamma + a_0 \frac{\Delta\tau}{\Delta t}\frac{\partial U}{\partial Q}\right)^n (Q^k - Q^{k-1}) = \Delta\tau\left[R(Q^{k-1}) - \frac{1}{\Delta t}(a_0 U^{k-1} - a_1 U^n + a_2 U^{n-1})\right],$$

where $U^{k-1} = (\partial U/\partial Q)^n Q^{k-1}$. At the pseudo-time iteration k, the variables are found by

$$Q^k = Q^{k-1} + \Delta\tau K^{-1}\left[R(Q^{k-1}) + \frac{1}{\Delta t}(a_0 U^{k-1} - a_1 U^n + a_2 U^{n-1})\right].$$

The matrix K^{-1} is given by

$$K^{-1} = \left(\Gamma + m\frac{\partial U}{\partial Q}\right)^n,$$

where $m = a_0\Delta\tau/\Delta t$.

If the second-order Runge–Kutta scheme is used for discretization with respect to pseudo-time, the inner loop is written as

$$Q^* = Q^{k-1} + \Delta\tau K^{-1}\left[R(Q^{k-1}) - \frac{1}{\Delta t}(a_0 U^{k-1} - a_1 U^n + a_2 U^{n-1})\right];$$

$$Q^k = Q^{k-1} + \Delta\tau K^{-1}\left[R(Q^*) - \frac{1}{\Delta t}(a_0 U^* - a_1 U^n + a_2 U^{n-1})\right].$$

The structure of the matrix K does not change in this case.

At the pseudo-time iteration $k-1$, the solution at the current time layer $(Q^{k-1} = Q^n)$ is used as the initial approximation. The inner loop on iterations k is executed, and the solution Q^k is found. The inner iterations at each physical time layer ensure that the derivative with respect to the pseudo-time tends to zero. In the inner iteration loop, the vectors Q^{n-1} and Q^n are assumed to be constant, and the vector Q^k is computed based on the solution U_k obtained at the iteration k. The iterative process is terminated when a predefined number of iterations is reached or the convergence criterion (the minimum norm of the residual does not exceed the predefined error) is fulfilled. At the next time layer $n+1$, the solution at iteration $k(Q^{n+1} = Q^k)$ is used as the solution.

The step Δt with respect to the physical time is chosen based on the accuracy requirements, and the pseudo-time step $\Delta\tau$ is determined by the Courant–Friedrichs–Lewy (CFL) condition.

4.3.2 Implicit Scheme

The use of the first-order implicit Euler scheme for the discretization of Eq. (7) yields the equation

$$\frac{U_i^{n+1} - U_i^n}{\Delta t} = R(U^{n+1}). \tag{12}$$

The index n refers to the time layer. The implementation of this approach requires the computation of the derivative $\partial R/\partial U$ in terms of the known values of the solution vector.

The differential term with respect to pseudo-time is added to the left-hand side of Eq. (7)

$$\Gamma \frac{\partial U}{\partial \tau} = R_*(U), \tag{13}$$

where

$$R_*(U) = R(U) - \frac{\partial U}{\partial t}.$$

The first-order implicit Euler scheme is used for the discretization with respect to the pseudo-time, while the implicit three-layer scheme BSD-2 is used for the discretization with respect to the physical time. Dropping the index $n+1$, which refers to the time layer, gives

$$\Gamma \frac{U^{m+1} - U^m}{\Delta \tau} = R^{m+1}(Q) - \frac{1.5Q^{m+1} - 2Q^n + 0.5Q^{n-1}}{\Delta t}. \tag{14}$$

The index n refers to the physical time (outer iterations), and m corresponds to the pseudo-time layer (inner iterations).

The expansion of the residual at the node i of the layer $m+1$ into Taylor series with respect to the pseudo-time in a neighborhood of the solution at iteration m is

$$R^{m+1} = R^m + \frac{\partial R}{\partial U} \Delta U, \tag{15}$$

where $\Delta U = U^{m+1} - U^m$. Taking into account the linearization of residual (15), Eq. (14) can be written as

$$\left[\left(\frac{\Gamma}{\Delta \tau} + \frac{I}{\Delta t} \right) + \frac{\partial R^m}{\partial U} \right] \Delta U^m = R^m - \frac{1.5Q^m - 2Q^n + 0.5Q^{n-1}}{\Delta t}.$$

The system of difference equations is solved using the GMRES method. At $m = 1$, the solution to this problem yields the vector U^{n+1} (in this case, $R_* \to 0$). At $\Delta \tau = \infty$, the steady-state solution to the problem is obtained in one step.

The iterative process with respect to pseudo-time is terminated when a predefined number of iterations is reached or the convergence criterion is fulfilled. When the convergence with respect to the inner steps is reached, the transition to the next step with respect to the physical time is made ($Q^{n-1} = Q^n$, $Q^n = Q^m$).

4.3.3 Residual Smoothing

To increase the convergence rate with respect to pseudo-time, residual smoothing is used (Zhao et al. 2002). The residual at the mesh node (at the center of the control volume) is replaced with the smoothed or weighted average of the residuals in the adjacent mesh nodes. To increase the local pseudo-time step, the average residual is computed implicitly.

At the node k of the unstructured mesh, the average residual is found by solving the equation (Zhao et al. 2002)

$$\overline{R}_k = R_k + \varepsilon \nabla^2 \overline{R}_k, \tag{16}$$

where R is the initial residual and $\overline{R}$ is the smoothed residual. The smoothing parameter is found by the formula (Zhao et al. 2002)

$$\varepsilon = \max\left\{\frac{1}{4}\left[\left(\frac{\text{CFL}}{\text{CFL}^*}\right)^2 - 1\right], 0\right\},$$

where CFL^* is the maximum Courant number of the basic scheme. Equation (16) is solved using the iterative Jacobi method

$$\overline{R}_k^{(m)} = R_k^{(0)} + \varepsilon \sum_{i=1}^{N_k} \left[\overline{R}_i^{(m)} - \overline{R}_k^{(m)}\right].$$

Rearranging the terms yields

$$\overline{R}_k^{(m)} = \left[R_k^{(0)} + \varepsilon \sum_{k=1}^{N_k} \overline{R}_i^{(m-1,m)}\right] (1 + \varepsilon N_k)^{-1},$$

where N_k is the number of nodes adjacent to the node k.

4.4 Numerical Examples

The capabilities of the preconditioning method are demonstrated by computation of the free convective flow in the gap between the coaxial cylinders. In the computational algorithm, the mesh was stored using information about the connections between the control volumes (the connectivity matrix). To simplify the code parallelization, explicit difference schemes were used. The computations were performed on a general purpose graphics processing unit (Emelyanov et al. 2015).

In the computations of free convective flows in the incompressible fluid model, the Boussinesq approximation is widely used (in this case, the dynamic and heat problems turn out to be interconnected). In the Boussinesq approximation, the linear dependence of the density on temperature $\rho(T) = \rho_0[1 - \beta(T - T_0)]$ is used, where β is the coefficient of volumetric expansion, and ρ_0 is the density at a reference temperature T_0. The Boussinesq approximation adequately describes the flow under mass forces when the difference of temperatures of the cold and hot walls does not exceed 30° (Gray and Giorgini 1976).

In the general case, free convective flows are modeled using the Navier–Stokes equation (7) with the source term in the momentum equation, which takes into account the action of potential mass forces. Free convective flows have low velocities due to which they can be computed using preconditioning methods (Volkov 2009).

In Weiss and Smith (1995), Fu et al. (2009), free convective flows are simulated based on the full Navier–Stokes equations. In Weiss and Smith (1995), the preconditioning method is used to simulate free convection between concentric circles (the Rayleigh number is 4.7×10^4, and the ratio of the diameters of the outer and inner cylinders is 2.6). The results are compared with the results of physical and numerical experiments presented in Kuehn and Goldstein (1976, 1978) (the Rayleigh number varies in the range 2.11×10^4 through 9.76×10^5, and the computations are performed in the Boussinesq approximation). The unsteady flow pattern occurs at Rayleigh numbers greater than 10^5. Three-dimensional computations within the Boussinesq approximation were performed in Desai and Vafai (1994). A comparison of various turbulence models can be found in Char and Hsu (1998). The presence of side walls results in the decrease of the heat transfer coefficient, which mainly is manifested at large Rayleigh numbers.

Consider natural convection in a horizontal annulus (Weiss and Smith 1995). The computational domain is depicted in Fig. 11. The diameters of the inner and outer cylinders are $d_i = 3.56$ cm and $d_o = 9.25$ cm, and their length is $l = 20.32$ cm.

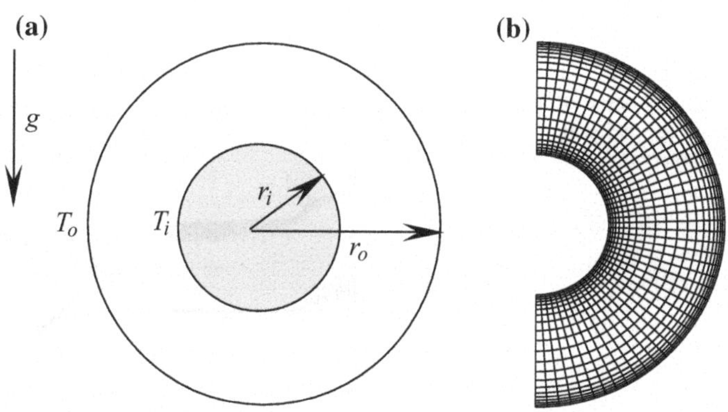

Fig. 11 Computational domain (**a**) and mesh (**b**)

The flow structure depends on the ratio of the diameters of the outer and inner cylinders ($d_o/d_i = 2.6$) and on the Rayleigh number

$$Ra_L = \frac{g\beta(T_i - T_o)L^3}{\nu\alpha},$$

where $L = r_o - r_i$. In CFD computations, the Rayleigh number varied in the range between 10^2 and 10^5. In the basic example used in comparison with the experimental data presented in Kuehn and Goldstein (1976, 1978), it was $Ra_L = 4.7 \times 10^4$ (the temperature difference was $\Delta T = 26.3$ K, and the temperatures of the inner and outer cylinders were 321.38 and 295.08 K). The computation results were processed in terms of the dimensionless temperature and dimensionless radius

$$\varphi = \frac{T - T_o}{T_i - T_o}, \quad \eta = \frac{r - r_i}{r_o - r_i}.$$

The computations were performed on a mesh consisting of 980 cells condensing near the boundaries of the inner and outer cylinders to match the boundary layer. The mesh in a half of the computational domain is shown in Fig. 11b (the mesh is stored in an unstructured form).

In the computations without preconditioning, the stagnation of the iterative process is observed, i.e., the residual remains almost the same as the number of iteration steps increases, which is clearly seen in Fig. 12a. The use of preconditioning considerably speeds up the convergence rate, and the residual decreases by six orders of magnitude in 10^4 iteration steps as is shown in Fig. 12b. Lines 1 correspond to pressure residual, lines 2 correspond to velocity residual, and lines 3 correspond to temperature residual.

The temperature contours, the flow direction, and the current lines in the middle plane between two horizontal cylinders are shown in Figs. 13, 14, 15, and 16 for various Rayleigh numbers.

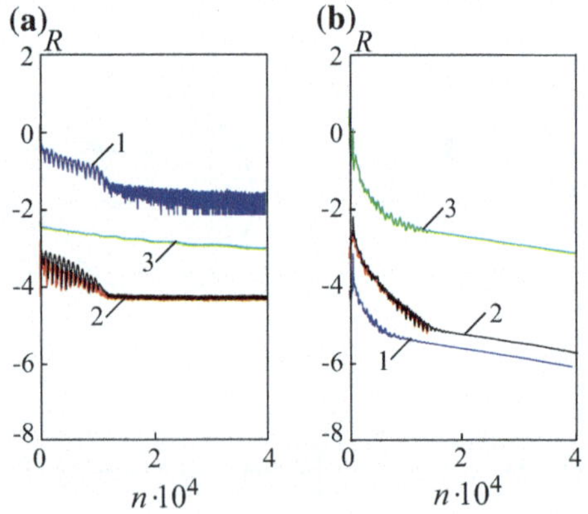

Fig. 12 Convergence of the iterative process based on the Roe scheme (**a**) and with the use of the preconditioning option (**b**)

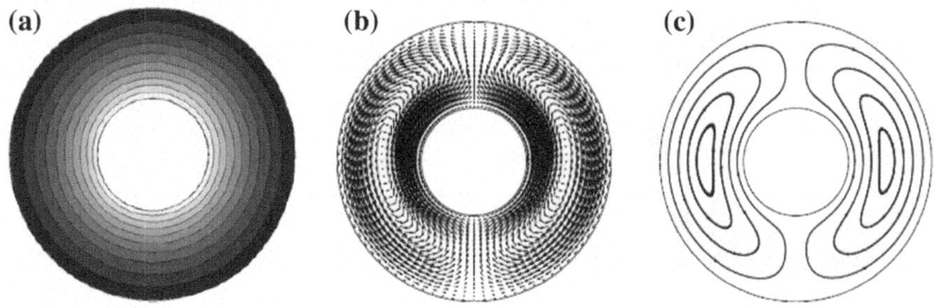

Fig. 13 Temperature contours (**a**), vector field (**b**), and streamlines (**c**) at $Ra = 10^2$

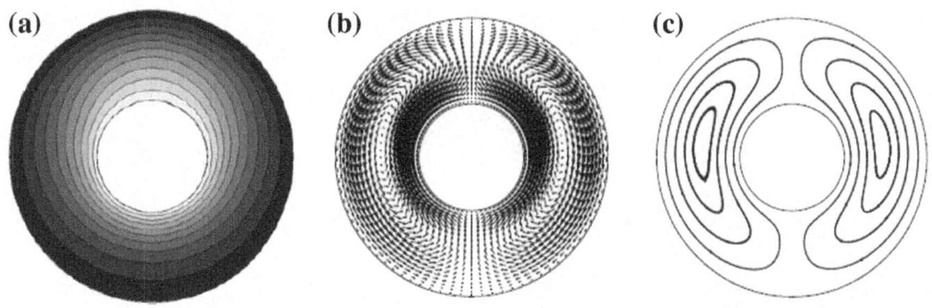

Fig. 14 Temperature contours (**a**), vector field (**b**), and streamlines (**c**) at $Ra = 10^3$

At $Ra = 10^2$, the heat exchange is due to thermal conduction, and the temperature contours are concentric circles (Fig. 13). The increase in the Rayleigh number to 10^3 results in the increasing role of the convective heat transfer, and the temperature contours take a slightly elongated shape in the vertical direction instead of concentric circles (Fig. 14). The further increase in the Rayleigh number to 10^4 makes the convective transfer the major factor in determining the flow and heat exchange pattern (Fig. 15). The increase in the Rayleigh number also changes the position of the recirculation zones as a result of which their centers move upward in the vertical direction. At $Ra = 10^5$ (Fig. 16), the depth of the boundary layer on the cylinder surface increases, and the temperature contours become more elongated in the vertical direction.

The velocity vector field and the temperature level lines in the basic computation example (at $Ra = 4.7 \times 10^4$) are shown in Fig. 17.

At $Ra = 4.7 \times 10^4$, the distributions of the dimensionless temperature as a function of the dimensionless radius in the symmetry planes arranged at the angles of 0°, 90°, and 180° to the vertical plane are shown in Fig. 18 for the numerical and physical experiments described in Weiss and Smith (1995), Kuehn and Goldstein (1976). A good agreement of the computational and experimental data in various planes is observed.

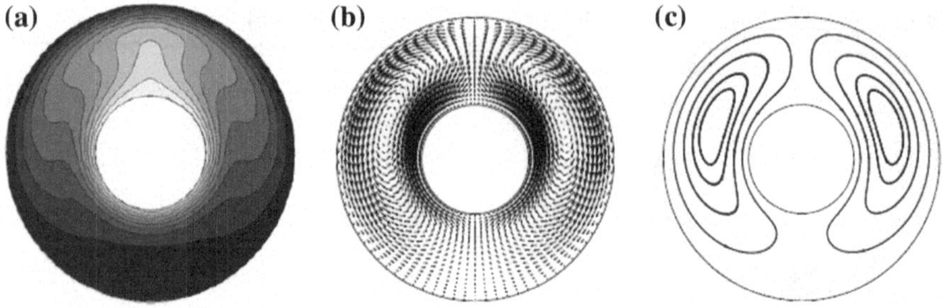

Fig. 15 Temperature contours (**a**), vector field (**b**), and streamlines (**c**) at $Ra = 10^4$

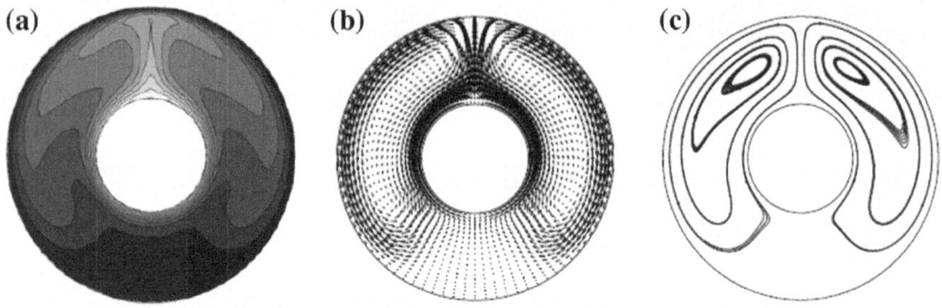

Fig. 16 Temperature contours (**a**), vector field (**b**), and streamlines (**c**) at $Ra = 10^5$

Fig. 17 Vector field (left) and temperature contours (right)

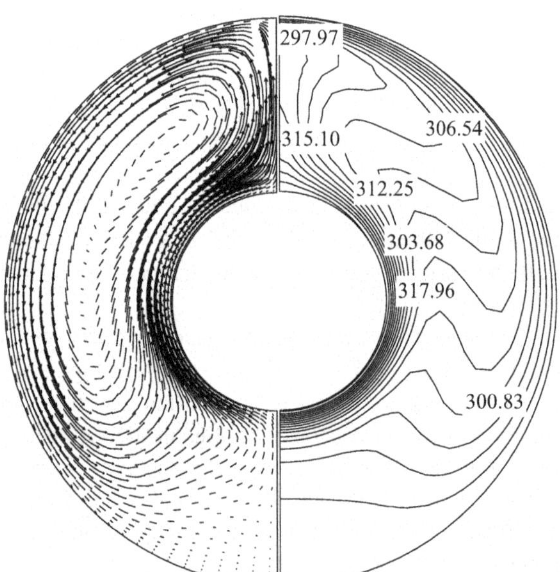

Fig. 18 Distributions of temperature in the symmetry planes tilted 0° (line 1), 90° (line 2), and 180° (line 3) with respect to the vertical plane compared with the data of (Kuehn and Goldstein 1976) (symbols •)

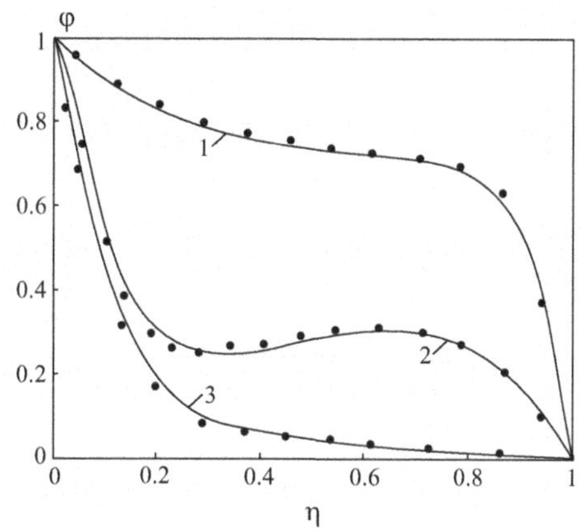

Fig. 19 Distributions of temperature at $Ra = 10^2$ (1), 10^3 (2), 10^4 (3), and 10^5 (4)

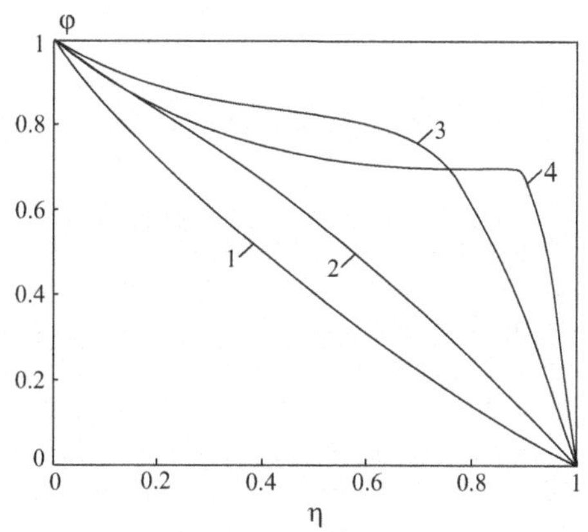

The temperature decreases in the domain between the walls of the outer and the inner cylinders, which corresponds to the difference of the fixed temperatures of the two cylinder surfaces. The warm air rises upward and moves from the wall of the inner cylinder to the outer one and sinks to the lower part of the annulus upon cooling. The air recirculates inside the annulus due to natural convection. The greatest speed is observed near the cylinder walls in the upper part.

The distributions of temperature in the gap between the cylinders in the horizontal plane are shown in Fig. 19 at various Rayleigh numbers. At $Ra = 10^2$ (line 1), the temperature in the gap between the cylinders changes almost linearly (the heat transfer is mainly due to thermal conduction). The increase in the Rayleigh number up to 10^5 (line 4) results in a large temperature gradient on the outer cylinder surface.

The local Nusselt numbers on the inner and outer cylinder surfaces are found by

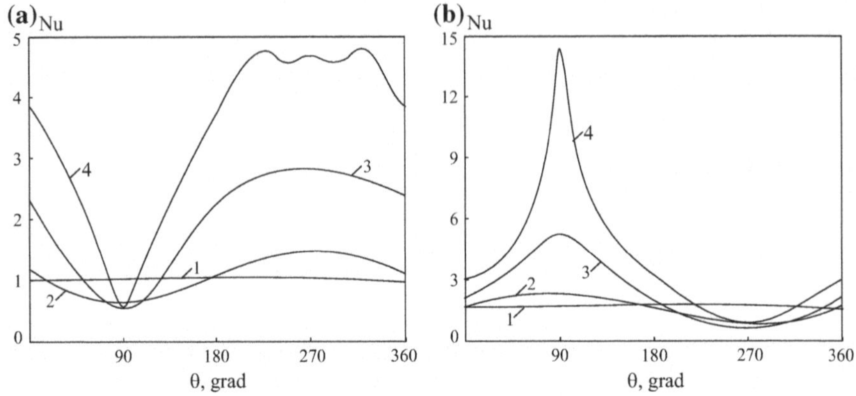

Fig. 20 Distributions of the local Nusselt number over the surface of the inner (**a**) and outer (**b**) cylinders at $Ra = 10^2$ (1), 10^3 (2), 10^4 (3), and 10^5 (4)

$$Nu_{\rm i} = r_{\rm i} \ln\left(\frac{r_{\rm o}}{r_{\rm i}}\right) \frac{\partial T}{\partial r}\bigg|_{r=r_{\rm i}}, \quad Nu_{\rm o} = r_{\rm o} \ln\left(\frac{r_{\rm o}}{r_{\rm i}}\right) \frac{\partial T}{\partial r}\bigg|_{r=r_{\rm o}}.$$

The average Nusselt number (denoted by angle brackets) is the arithmetic mean of the Nusselt numbers corresponding to the inner and outer cylinders.

The distributions of the local Nusselt numbers over the inner and outer cylinder surfaces are shown in Fig. 20 for various Rayleigh numbers. In the horizontal plane, the Nusselt number has the minimum on the inner cylinder surface, and it has the maximum on the outer cylinder surface. It is remarkable that at $Ra = 10^5$, the Nusselt number varies non-monotonically on the surface of the inner cylinder. The local maximum of the Nusselt number is in the planes tilted 220° and 320° with respect to the vertical, and the local minimum of the Nusselt number is on the outer cylinder surface in the plane tilted 270° with respect to the vertical.

The variation of the average Nusselt number in time is illustrated in Fig. 21. The dashed lines correspond to the inner cylinder surface, and the solid lines correspond to the outer cylinder surface. The increase in the Nusselt number results in the decrease of the time needed to achieve a steady-state solution.

Figures 22 and 23 illustrate the comparison of the computed distributions of the Nusselt number with the data of physical experiments presented in Kuehn and Goldstein (1976). There is a good agreement between the numerical and experimental data both in respect to the local (the local Nusselt number) and the integral (average Nusselt number) heat exchange characteristics.

The experimental and numerical results presented in Kuehn and Goldstein (1976, 1978) were analyzed using the equivalent thermal conductivity method, which is defined as the ratio of the measured heat transfer in the annulus to the heat transfer due exclusively to the convective heat exchange. For the conditions of the problem at $Ra = 4.7 \times 10^4$, the equivalent thermal conductivity is 2.58 W/m

Fig. 21 Variation of the average Nusselt number in time at $Ra = 10^3$ (1), 10^4 (2), and 10^5 (3)

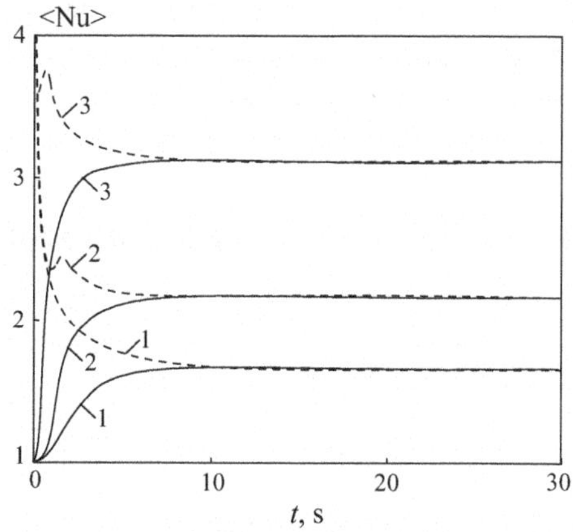

Fig. 22 Distributions of the local Nusselt number over the surface of the inner (1) and outer (2) cylinders compared with data from (Kuehn and Goldstein 1976) (symbols •) at $Ra = 4.7 \times 10^4$

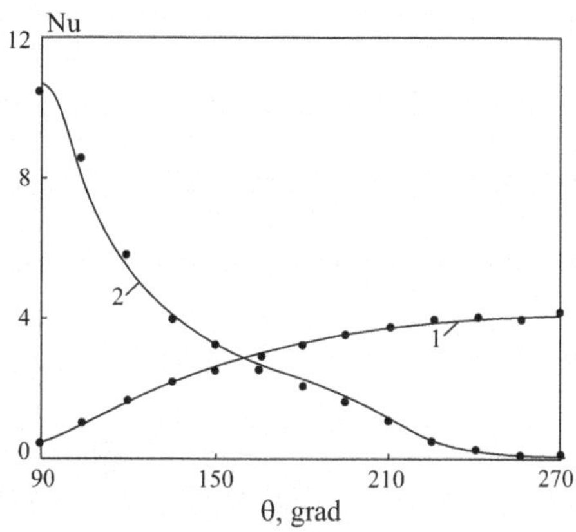

Fig. 23 Distributions of the average Nusselt number in the computations (solid line) and physical experiment (Kuehn and Goldstein 1976) (symbols •)

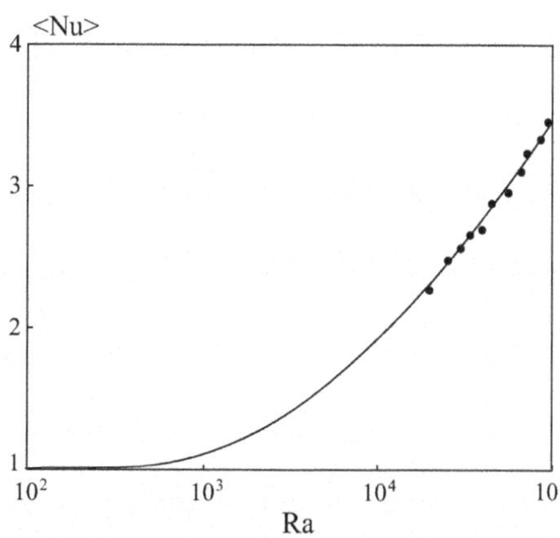

(for the unit length of the cylinder) (Kuehn and Goldstein 1976). The numerically obtained thermal conductivity is 2.54 W/m, which gives the error of 1.6% compared with the physical experiment.

5 Geometric and Algebraic Multigrid Techniques

The construction of prolongation and interpolation operators, as well as mesh levels of various resolutions, is discussed.

5.1 *Geometric Methods*

Geometric multigrid (GMG) methods are the classical multigrid techniques proposed in Fedorenko (1961). The system of difference equations is solved on a sequence of meshes of different resolution (Fig. 24a). The method used to construct the sequence of meshes and its adaptation to the problem (inviscid or viscous, laminar or turbulent) predetermines to a large extent the success of the application of the multigrid approach (Wesseling and Oosterlee 2001; Brannick et al. 2006). A proper choice of the prolongation and restriction operators makes it possible to smooth the high-frequency modes of the solution error when going to a coarser mesh level (Fig. 24b).

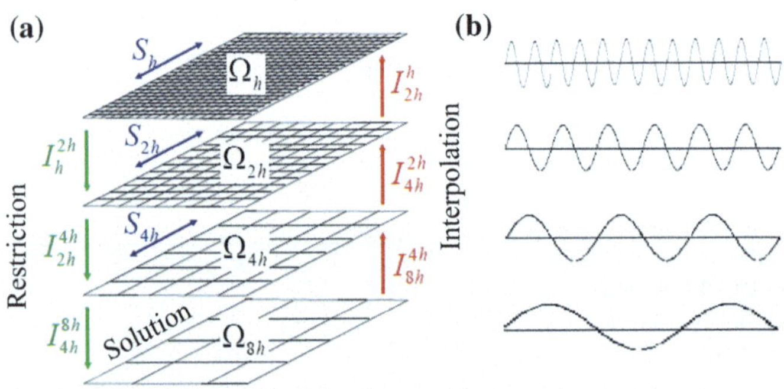

Fig. 24 Geometric multigrid technique

5.1.1 Full Approximation Scheme

Consider the system of difference equations

$$N(Q) = f, \tag{17}$$

where N is a discrete operator and Q is the vector of conservative variables. The vector f is composed of the coefficients that depend on discretization and of the known values of the solution vector specified by the boundary conditions.

The iterative procedure for solving the system of Eq. (17) is written as

$$Q^{n+1} = Q^n + J^{-1}[f - N(Q^n)] \quad (n = 1, 2, \ldots),$$

where J is the Jacobian and $R = f - N(Q^n)$ is the residual. The superscript n indicates the iteration number.

The multigrid technology designed for solving systems of difference Eq. (17) is based on computations on a sequence of nested meshes $h_1 \supset \cdots \supset h_M$, which generate a sequence of finite-dimensional spaces $V_{h_1} \supset \cdots \supset V_{h_M}$, and on the sequence of operator equations

$$N^k(Q^k) = f^k, \quad Q^k \in V_{h_k} \quad (k = 1, \ldots, M),$$

where M is the number of mesh levels. The coarse mesh solution is used to correct the solution on the fine mesh.

In discrete form on the fine mesh

$$N^h(\widehat{Q}^h) = f^h, \tag{18}$$

where $\widehat{Q}^h$ is the exact solution to the discrete system. By using Q^h as an initial approximation and determining the solution error by $E^h = \widehat{Q}^h - Q^h$, Eq. (18) is rewritten in the form

$$N^h(Q^h + E^h) = f^h. \tag{19}$$

Subtract $N^h(Q^h)$ from both parts of Eq. (19) to obtain

$$N^h(Q^h + E^h) - N^h(Q^h) = f^h - N^h(Q^h) = R^h(Q^h).$$

The residual and the solution are restricted to the coarse mesh

$$N^H(E^H) = I_h^H R^h(Q^h);$$
$$N^H(I_h^H Q^h + E^H) - N^H(I_h^H Q^h) = I_h^H[f^h - N^h(Q^h)].$$

Here, I_h^H and I_H^h are the restriction operators from the fine mesh to the coarse one $(H \to h)$ and the prolongation operator from the coarse mesh to the fine one $(h \to H)$. In particular, $Q^H = I_h^H R^h$ and $Q^h = I_H^h R^H$.

On the coarse mesh

$$f^H = I_h^H \left[f^h - N^h(Q^h) \right] + N^H(I_h^H Q^h).$$

To solve the system of equations $N^H(Q^H) = f^H$ on the coarse mesh, n_c smoothing iterations are performed.

The correction made when going from the fine mesh to the coarse one is

$$T_h^H = N^H(I_h^H Q^h) - I_h^H N^h(Q^h).$$

The coarse mesh equation is corrected in order to make the coarse mesh solution coincide with the fine mesh solution.

5.1.2 Prolongation and Restriction Operators

At the step of prolongation from the coarse mesh to the fine one, the linear interpolation is used to reconstruct the solution

$$\Delta Q_i^h = \Delta Q_j^H + \left(x_i^h - x_j^H \right) \cdot \nabla \left(\Delta Q^H \right)_j \quad \text{for} \, \forall \, i \in K_j, \tag{20}$$

where ΔQ is the correction of the solution. Indices h and H are related to fine and coarse meshes.

At the step of restriction from the fine mesh to the coarse one, the mean value residual is calculated

$$R_j^H = \sum_{i \in K_j} V_i^h R_i^h \Big/ \sum_{i \in K_j} V_i^h. \tag{21}$$

The averaging over volume assumes that it holds that

$$V_j^H = \sum_{i \in K_j} V_i^h.$$

Near the boundary of the computation domain (near a wall), it can happen that

$$V_j^H > \sum_{i \in K_j} V_i^h.$$

Condition (21) is replaced with the modified relation

$$R_j^H = \sum_{i \in K_j} V_i^h R_i^h / \max\left(V_j^H, \sum_{i \in K_j} V_i^h \right). \tag{22}$$

The summation in (22) is over all nodes i of the mesh Ω^h, which, being joined, form the control volume V_i constructed near the node $j \in \Omega^H$. The set of nodes has the form $K_j = \{ i \in \Omega^h : \cup V_i^h = V_i^H \}$, and $\sum_{i \in K_j} V_i^h = V_j^H$.

5.1.3 Smoothing Procedure

The smoothing method that damps the high-frequency error modes seems to be the component of the multigrid method that most strongly depends on the type of the problem to be solved. Linear iterative methods (the Jacobi and the Gauss–Seidel method, incomplete factorization method) (Trottenberg et al. 2001) and the multistep Runge–Kutta methods (Jameson 1983) are conventional smoothing methods.

When steady flows are simulated using the time relaxation method, the multigrid method is used to speed up the convergence in the integration with respect to the pseudo-time; as the smoothing procedure, Runge–Kutta schemes are widely used (Jameson 1983). Rather than solve the nonlinear Eq. (17), the ordinary differential equation is solved

$$\frac{\mathrm{d}Q}{\mathrm{d}t} = N(Q) - f.$$

The multistage Runge–Kutta method has the form

$$Q^{(k)} = Q^{(0)} - c_k \tau N[Q^{(k-1)}] + c_k \tau f \quad (k = 1, \ldots, p).$$

Here, $Q^n = Q^{(0)}$ and $Q^{n+1} = Q^{(p)}$, and τ is the time integration step. The coefficients of the difference scheme, c_k, are chosen so as to ensure the effective smoothing rather than from the accuracy and stability consideration of the numerical solution (Jameson 1995).

5.1.4 Multigrid Cycle

The implementation of the multigrid method is reduced to the execution of the following sequence of steps.

1. Using the smoothing method, μ_1 approximations of the solution on the mesh h are made (pre-smoothing)

$$\widehat{Q}^h = Q^h + (J^h)^{-1}[f^h - N^h(Q^h)].$$

For simplicity, the superscript n indicating the iteration number is missed.

2. The residual $R^h = f^h - N^h(Q^h) \in V_h$ is projected onto the space V_H, therefore $R^H = I_h^H R^h$.
3. The coarse mesh solution $N^H(Q^H) = N^H(I_h^H Q^h) + R^H$ is found. To this end, γ recursive calls and n_c iterations on the coarsest mesh are made for smoothing.
4. The error $E^H = Q^H - I_h^H Q^h$ is interpolated to the fine mesh, and the solution on this mesh is corrected

$$\widehat{Q}^h = Q^h + I_H^h(Q^H - I_h^H Q^h).$$

5. To suppress the interpolation error, μ_2 approximations of the solution on the fine mesh are made (post-smoothing).

The iterations begin from the highest resolution mesh (level 1).

5.1.5 Sequence of Meshes

The sequence of nested meshes is constructed using the edge collapsing algorithm (Crumpton et al. 1997; Muller and Giles 1998). Figure 25 illustrates the edge collapsing procedure. Each edge is assigned its length, which is multiplied by the edge growth factor when going to the next mesh level (e.g., the growth factor can be equal to two). In the boundary layer, cells are collapsed in the direction of the shortest edge only if only its length is not less than a threshold, which enables preserving the mesh topology in the near wall domain. A constraint on the edge lengths is checked. The list of cells is composed, the maximum acceptable angle between two edges (e.g. 135°) is fixed, and the geometric criterion is checked (cells including a collapsing edge are examined).

The mesh construction procedure is recursive, and it can be implemented relatively easily. The list of all mesh cells is created and sorted by their size. As cells are collapsed, they are removed from this list. The algorithm stops when the cell list becomes empty. When the coarse mesh is constructed, it is checked that its volume is positive (the maximum angle between its edges) and that the edge length doubles

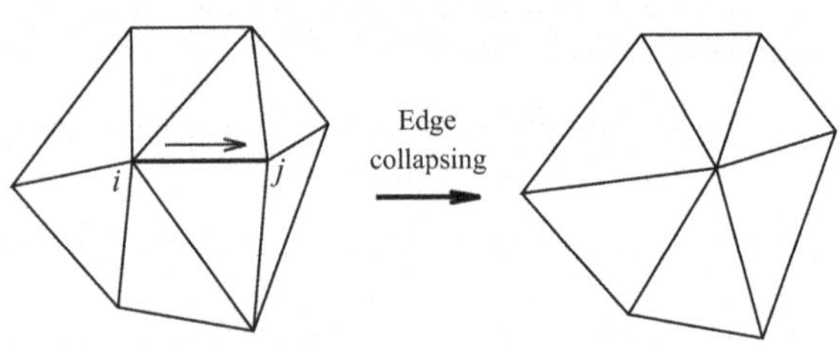

Fig. 25 Edge collapsing in the direction of the shortest edge

when going from the finer mesh to the coarser one. The mesh coarsening method is implemented in a special program module that is called by the preprocessor when the computation model is prepared (Volkov 2009).

To construct a mesh suitable for simulating viscous flows in the boundary layer, when cells flattened cells stretched along the wall are used, the directional mesh coarsening in the direction of the shortest cell edge is used (Muller and Giles 1998). The modified approach preserves the part of the mesh near the wall (e.g., near the airfoil). The loss of mesh regularity away from the airfoil (in the inviscid part of the flow) relatively weakly affects the quality of the numerical solution. The proposed approach reduces the degree of cell elongation of the coarse mesh in the boundary layer, which improves the convergence rate (Mavriplis 2002).

5.2 *Algebraic Methods*

The implementation of the algebraic multigrid method does not require information about the problem geometry. It uses only the structure of the coefficient matrix of the system of difference equations (Fig. 26). The implementation of the algebraic multigrid method consists of the mesh (matrix) formation phase on different levels and the solution phase (Cleary et al. 2000).

5.2.1 Basic Ideas

Consider the system of difference equations

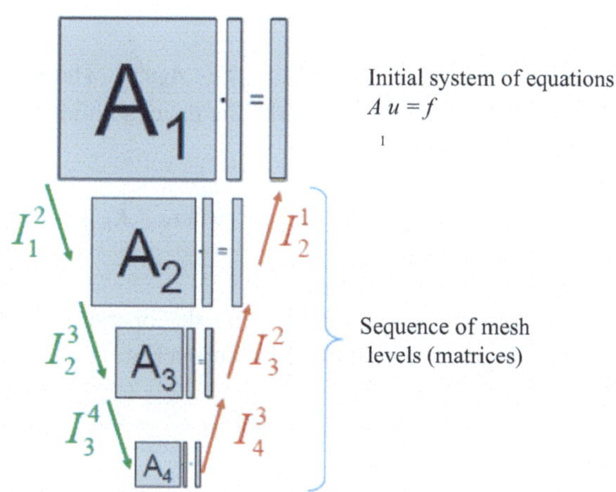

Fig. 26 Algebraic multigrid technique

$$A_h u^h = f^h \quad \text{or} \quad \sum_{j \in \Omega^h} a_{ij}^h u_j^h = f_i^h \quad (i \in \Omega^h). \tag{23}$$

The matrix A of size $n \times n$ in (23) consists of the coefficients a_{ij} $(i, j = 1, \ldots, n)$, determined by discretization. The vector u of length n consists of the unknown node values of the function to be found. The vector f of length n is composed of the coefficients obtained in the course of discretization and of the given values of the solution vector specified by the boundary conditions. The values of the unknown function at the nodes of the mesh $\Omega = \{1, 2, \ldots, n\}$ are denoted by $u_i (i = 1, \ldots, n)$.

In contrast to the geometric method in which the set of nodes Ω^h is the highest resolution mesh, in the multigrid algebraic method $\Omega^h = \{1, 2, \ldots, n\}$ is treated as the set of indices of unknowns. The set Ω^H has a lower number of elements, and it corresponds to the coarse mesh.

To transfer the solution from the coarse mesh to the fine one, the interpolation operator I_H^h is used, and the operator of restriction to the coarse mesh level I_h^H is found from the condition obtained using the Galerkin method

$$A_H = I_h^H A_h I_H^h, \quad \text{where} \quad I_h^H = (I_H^h)^{\mathrm{T}}.$$

The fine mesh solution is found from the equation

$$u_{\text{new}}^h = u_{\text{old}}^h + I_H^h e^H. \tag{24}$$

With regard to (24), the correction of the solution on the coarse mesh e^H is found by solving the system of equations

$$A_H e^H = r^H \quad \text{or} \quad \sum_{j \in \Omega^H} a_{ij}^H e_j^H = r_i^H \quad (i \in \Omega^H),$$

where $r^H = I_h^H (r_{\text{old}}^h)$ and $r_{\text{old}}^h = f^h - A_h u_{\text{old}}^h$. The error of the numerical solution $e^h = u_*^h - u^h$ (the asterisk marks the exact solution) is found using the coarse mesh correction operator

$$e_{\text{new}}^h = K_{h,H} e_{\text{old}}^h, \quad \text{where} \quad K_{h,H} = I_h - I_H^h A_H^{-1} I_h^H A_h, \tag{25}$$

where I_h is the identity operator.

The solution is smoothed using the operator S_h. In the smoothing phase (the smoothed solution is marked by an overline)

$$u^h \to \overline{u}^h, \quad \text{where} \quad \overline{u}^h = S_h u^h + (I_h - S_h) A_h^{-1} f^h.$$

Smoothing the solution error yields $e^h \to \overline{e}^h$, where $\overline{e}^h = S_h e^h$.

If μ_1 and μ_2 iterations are used for the pre- and post-smoothing, the correction operator is written as

$$e^h_{\text{new}} = M_{h,H} e^h_{\text{old}}, \quad \text{where} \quad M_{h,H}(\mu_1, \mu_2) = S_h^{\mu_2} K_{h,H} S_h^{\mu_1}.$$

In the classical approach, the variables on the coarse mesh level are constructed as a subset of variables on the fine mesh (Ruge and Stuben 1987; Stuben 2001a, b). The set of fine mesh variables is split into two disjoint subsets $\Omega^h = C^h \cup F^h$, where C^h is the set of coarse mesh variables, and F^h is the set of fine mesh variables. The solution error $e^h = I^h_H e^H$ is interpolated by the rule

$$e^h_i = \left(I^h_H e^H\right)_i = \begin{cases} e^H_i, & \text{if} \quad i \in C^h, \\ \sum\limits_{k \in P^h_i} w^h_{ik} e^H_k, & \text{if} \quad i \in F^h, \end{cases} \tag{26}$$

where $P^h_i \subset C^h$ is the set of variables involved in the interpolation. When the solution is interpolated from the coarse mesh to the fine mesh, the solution error e^h_i is set to e^H_i if i is a C-variable, and it is set equal to a weighted sum of variables from the set P^h_i if i belongs to the subset F. The set P^h_i is a small subset of the set of C-variables that are close to the variable i, which guarantees that the matrix A_H is sparse. On the other hand, the set P^h_i contains a sufficiently large number of variables to which i is strongly coupled.

The convergence of the multigrid method depends on the variation of the norm of the solution error

$$\left\| e^h \right\|_{A_h} = \left(A_h e^h, e^h\right)^{1/2}.$$

Using the variational principle and μ pre-smoothing iterations (Stuben 2001a, b)

$$\left\| K_{h,H} S_h^{\mu} e^h \right\|_{A_h} = \min_{e^H} \left\| S_h^{\mu} e^h - I^h_H e^H \right\|_{A_h}. \tag{27}$$

The procedure described by (27) can be recursively extended to any number of mesh levels. The convergence rate depends on the choice of the relaxation procedure and interpolation operator.

5.2.2 Implementation Steps

The set of coarse mesh variables is split into the subsets of C-variables and F-variables. The mesh level Ω^k $(k = 1, 2, \ldots, M-1)$ is assigned the subsets C^k and F^k. Each mesh level is also assigned the mesh operators A^1, A^2, …, A^M (where $A^1 = A$), and also the interpolation operators $P^k = I^k_{k+1}$ and the restriction operators $R^k = I^{k+1}_k$, where $k = 1, 2, \ldots, M-1$. The restriction is the transposed interpolation operator $R^k = (P^k)^{\mathrm{T}}$. To construct the matrix of the system on the coarse mesh level, the Galerkin product is calculated, $R^k A^k P^k$. The error on each mesh level is smoothed using the smoothing operator S^k $(k = 1, 2, \ldots, M - 1)$. The coarsening procedure is repeated until the size of the system becomes sufficiently low for the system of difference equations to be solved by a direct method (e.g., by the Gauss elimination method).

The subsets C^k and F^k, as well as the interpolation, restriction, and smoothing operators, are constructed in the setup phase.

Upon the execution of the setup phase, the solution to the original equation is found by recursively applying the multigrid procedure. The number of smoothing iterations and the number of recursive calls of the method on each mesh level are specified. The sensitivity of the solution phase to the choice of the smoothing procedure is typically relatively low, and classical iterative methods are usually used.

5.2.3 Construction of Mesh Levels

By taking into account strong couplings between variables in the interpolation procedure, the convergence can be accelerated, while weak couplings have a negative effect on the convergence rate thus increasing the computational complexity. The diversity of available approaches is due to the search of a compromise between the overall effectiveness of the algorithm and its computational complexity.

In many cases, the vector and matrix terminology is replaced by the terms related to fictitious meshes. The nodes of these meshes are interpreted as the nodes of a directed graph related to the given matrix. The mesh node $i \in \Omega^h$ (associated with the variable u_i^h) is connected to the variable $j \in \Omega^h$ if $a_{ij}^h \neq 0$. The set of variables adjacent to the variable i has the form

$$N_i^h = \left\{ j \in \Omega^h : j \neq i, \, a_{ij}^h \neq 0 \right\} \quad (i \in \Omega^h).$$

In the mesh terminology, the equation $A_h u^h = f^h$ is a fictitious mesh equation on the fine mesh Ω^h and is a fictitious mesh equation $A_H u^H = f^H$ on the coarse mesh $\Omega^H \subset \Omega^h$.

To choose the matrix coefficients that are used for constructing the coarse mesh level, the concepts of strong influence of a variable and strong coupling between variables are introduced (Stuben 2001a, b). The variable i strongly depends on the variable j, and the variable j strongly affects the variable i if the magnitude of the matrix element a_{ij} is greater than all the off-diagonal coefficients in the matrix (standard coarsening)

$$-a_{ij} \geqslant \theta \max_{k \neq i} \{-a_{ik}\}. \tag{28}$$

The parameter $0 < \theta \leqslant 1$ in (28) controls the number of strong couplings between variables (typically, $\theta = 0.25$). The set S_i is the set of all variables j that are strongly coupled with the variable i (the set of variables that strongly affect the variable i)

$$S_i = \left\{ j : j \neq i, -a_{ij} \geqslant \theta \max_{k \neq i} (-a_{ik}) \right\}.$$

The available theoretical approaches are applied to solving scalar second-order elliptic partial differential equations whose discretization yields a system of difference equations with an M-matrix (in this case, the convergence rate of the method is independent of the problem size). If there are both negative and positive off-diagonal elements, the following definitions are used:

$$a_{ij}^- = \begin{cases} a_{ij}, & \text{if} \quad a_{ij} < 0, \\ 0, & \text{if} \quad a_{ij} \geqslant 0, \end{cases} \qquad a_{ij}^+ = \begin{cases} 0, & \text{if} \quad a_{ij} \leqslant 0, \\ a_{ij}, & \text{if} \quad a_{ij} > 0. \end{cases}$$

In this case, two disjoint subsets of variables are

$$N_i^- = \{ j \in N_i : a_{ij}^h < 0 \}, \quad N_i^+ = \{ j \in N_i : a_{ij}^h > 0 \}.$$

In practice, C/F splitting is constructed in such a way that the set of C-variables is approximately the maximally independent set (within the set C, the variables have no strong couplings between themselves), and the F-variables are surrounded by interpolatory C-variables (Yang 2006).

5.2.4 Interpolation

For simplicity, the superscript h is missed, and the relation (26) is replaced by

$$e_i = \sum_{k \in P_i} w_{ik} e_k \quad (i \in F). \tag{29}$$

After several relaxation steps, the scaled residual of the solution becomes much less than the solution error $|r_i| \ll a_{ii} |e_i|$. The local approximation of an algebraically smooth error such that $\mathrm{S}e = e$ and $|r_i| \ll a_{ii} |e_i|$ is

$$e_i = -\frac{1}{a_{ii}} \sum_{j \in N_i} a_{ij} e_j \quad (i \in \Omega). \tag{30}$$

In the implementation of the method, one must construct a proper C/F splitting and determine the set of interpolatory variables $P_i \subset C (i \in F)$ and weighting coefficients w_{ik} such that relation (29) gives an acceptable approximation for any algebraically smooth error.

The exact interpolation procedure is given by formula (30) in which $P_i = N_i$ and $w_{ik} = -a_{ik}/a_{ii}$. This approach requires that a C/F splitting be constructed such that, for each variable $i \in F$, all the variables adjacent to i are contained in the set C and result in the construction of a direct method for solving the system of difference equations (Stuben 2001a, b). However, the direct methods are computationally inefficient. To design an efficient computation procedure, small subsets P_i are needed (as small as possible) to obtain a sparse Galerkin operator and achieve a high convergence rate.

In the case of direct interpolation $P_i \subseteq N_i$, and the interpolation operator is constructed based on relation (30) given the subset of variables that are not involved in the interpolation (the part of the sum corresponding to the algebraically smooth error for which $j \in N_i \backslash P_i$).

For M-matrices, the interpolation procedure is constructed based on the fact that the algebraically smooth solution error changes slowly in the direction of the strongest coupling between variables. For any variable i that is strongly coupled with variables in the set P_i, the best approximation for the algebraically smooth error satisfies the equation

$$\left(\sum_{k \in P_i} a_{ik}\right)^{-1} \sum_{k \in P_i} a_{ik} e_k = \left(\sum_{j \in N_i} a_{ij}\right)^{-1} \sum_{j \in N_i} a_{ij} e_j \quad (i \in \Omega).$$

Substitution into Eq. (30) results in the interpolation procedure (29) with positive weighting coefficients (Stuben 2001a, b)

$$w_{ik} = -\alpha_i \frac{a_{ik}}{a_{ii}}, \quad \text{where} \quad \alpha_i = \frac{\sum_{j \in N_i} a_{ij}}{\sum_{l \in P_i} a_{il}} \quad (i \in F, l \in P_i). \tag{31}$$

In practice, relation (31) implies that, for the constructed splitting C/F, each variable $i \in F$ has a fairly large number of variables in the subset C with which it is strongly coupled and which are used as the interpolatory variables P_i.

For matrices of general form, the coefficients a_{ij}^+ and a_{ij}^- and the subsets N_i^+ and N_i^- are used, while positive coefficients are added to the coefficients on the principal diagonal. The interpolation procedure is given by the formula

$$\widetilde{a}_{ii} e_i + \alpha_i \sum_{k \in P_i} a_{ik}^- e_k = 0, \tag{32}$$

where

$$\widetilde{a}_{ii} = a_{ii} + \sum_{j \in N_i} a_{ij}^+, \quad \alpha_i = \frac{\sum_{j \in N_i} a_{ij}^-}{\sum_{k \in P_i} a_{ik}^-}.$$

This approach yields the positive weighting coefficients

$$w_{ik} = -\alpha_i \frac{a_{ik}^-}{\widetilde{a}_{ii}} \quad (i \in F, k \in P_i).$$

In the general case, taking into account positive couplings in the construction of the interpolation procedure turns out to be more important than the account for negative couplings (Stuben 2001a, b) (for positive couplings, the solution error varies most slowly). For a variable $i \in F$ that has both negative and positive couplings, it is assumed that $N_i^- \neq \emptyset$ and $N_i^+ \neq \emptyset$, while the C/F splitting is such that at least one coupling of any sign belongs to the subset C. Two sets of interpolatory variables such that $\emptyset = P_i^- \subseteq C \cap N_i^-$ and $\emptyset = P_i^+ \subseteq C \cap N_i^+$ are chosen.

For each variable $i \in F$, a set of interpolatory variables $P_i = N_i$ is specified, which gives

$$a_{ii} e_i + \sum_{j \in N_i} a_{ij} e_j = 0.$$

Using $P_i = P_i^- \cup P_i^+$ gives

$$a_{ii} e_i + \alpha_i \sum_{k \in P_i} a_{ik}^- e_k + \beta_i \sum_{k \in P_i} a_{ik}^+ e_k = 0. \tag{33}$$

The error is found from the equation

$$e_i = \sum_{k \in P_i} w_{ik} e_k,$$

where

$$\alpha_i = \frac{\sum_{j \in N_i} a_{ij}^-}{\sum_{k \in P_i} a_{ik}^-}, \quad \beta_i = \frac{\sum_{j \in N_i} a_{ij}^+}{\sum_{k \in P_i} a_{ik}^+}.$$

The weighting coefficients are found from the formulas

$$w_{ik} = \begin{cases} -\alpha_i a_{ik}/a_{ii} & \text{for} \quad k \in P_i^-, \\ -\beta_i a_{ik}/a_{ii} & \text{for} \quad k \in P_i^+. \end{cases}$$

In this case $w_{ik} > 0$, if $k \in P_i^-$ and $w_{ik} < 0$ if $k \in P_i^+$. If $N_i^+ = \emptyset$ or $N_i^- = \emptyset$, then the specifications are changed by setting $P_i^+ = \emptyset$, $\beta_i = 0$ and $P_i^- = \emptyset$, $\alpha_i = \emptyset$. If there are positive elements, they are added to the diagonal elements.

To improve the interpolation procedure, it is assumed that there is a C/F splitting such that, for each variable $i \in F$, the set of interpolatory variables $P_i \subseteq N_i \cap C$ is given. Rather than directly approximate the noninterpolatory part of the equation for the variable in (30), the solution errors e_j for all variables $j \notin P_i$ that are strongly coupled with the variable i in equation j are removed. The standard interpolation procedure results in expanding the set of interpolatory variables (the sparseness of the Galerkin operator is decreased).

5.2.5 Smoothing

The application of the general smoothing procedure gives

$$e_{n+1} = Se_n \quad \text{or} \quad u_{n+1} = Su_n + (I - S)A^{-1}f, \tag{34}$$

where $e_n = u_n - u$ is the error. The matrix $S = I - Q^{-1}A$, where Q is the lower triangular part of the matrix A, is chosen as the smoothing operator. The iterative scheme used as the smoothing procedure is written as

$$u_{n+1} = u_n + Q^{-1}(f - Au_n). \tag{35}$$

The Gauss–Seidel method is widely used as the smoothing procedure.

5.3 *Efficiency Indicators*

The main factors characterizing the efficiency of a multigrid procedure implementation are the convergence factor and the complexity factor.

The convergence rate is the decrease of the residual norm in two adjacent cycles $\rho = \|r^{k+1}\|/\|r^k\|$. The convergence factor is obtained by averaging the convergence rate over all multigrid cycles (after the prescribed residual level has been reached) or over a given number of cycles if the prescribed residual level cannot be achieved. The asymptotic convergence factor is computed relative to the residual norm at the first multigrid cycle.

The complexity factor is defined as the number of operations executed at one iteration step and the required memory at this step. The operator complexity is defined as the ratio of the total number of nonzero elements in the matrices A^k on all mesh levels $k = 1, 2, \ldots, M$ to the number of nonzero elements in the matrix A^1 corresponding to the finest mesh. It shows the number of operations per a multigrid cycle in one step of the solution. As the operator complexity decreases, the computation time decreases as well.

The operator complexity is an indicator of the amount of memory needed for the implementation of the multigrid method

$$C_\mathrm{O} = \sum_k \frac{|A^{(k)}|}{|A^{(1)}|},$$

where $|A^{(k)}|$ is the number of nonzero coefficients at the level k.

The mesh complexity factor shows how quickly the mesh is coarsened

$$C_\mathrm{G} = \sum_k \frac{n_k}{n_1},$$

where n_k is the number of variables (mesh points or graph nodes) at the level k.

The cycle complexity is characterized by the parameter

$$C_\mathrm{C} = \sum_k \frac{|A^{(k)}|}{|A^{(1)}|} \mu_k \gamma^k,$$

where $\mu_k = \mu_1 + \mu_2$. The parameter γ corresponds to the type of the multigrid cycle ($\gamma = 1$ for V-cycle and $\gamma = 2$ for W-cycle). A combination of the convergence and complexity factors of a multigrid cycle gives an estimate of the amount of computational work per unit of error $C_\mathrm{W} = -C_\mathrm{C}/\log \rho$.

5.4 Comparative Characteristics

The standard multigrid techniques are used to solve the Euler and Navier–Stokes equations with different degrees of success. The main factors that complicate the application of these techniques are the nonlinearity of equations and their mixed type (the standard multigrid techniques are developed for elliptic equations). Other difficulties arise in the solution of singularly perturbed problems (Trottenberg et al. 2001).

The geometric multigrid techniques seem to be more appropriate for solving nonlinear problems because the nonlinearities of the given equation are transferred down the mesh hierarchy (from finer to coarser meshes).

The algebraic multigrid techniques are used in the domains where the geometric methods are impractical or run into severe difficulties (Trottenberg et al. 2001) (e.g., the solution of a system of difference equations obtained by the discretization of partial differential equations with discontinuous coefficients). They need considerable time for the setup phase. For this reason, they are not used for solving small size and medium size systems of equations or in the case of low-order discretization.

While the successful application of geometric methods considerably depends on the smoothing procedure, the algebraic methods typically employ a simple relaxation procedure (the Jacobi or Gauss–Seidel methods) and an operator-dependent interpolation that gives a good representation of the norm of the solution error (Wesseling 1992; Wesseling and Oosterlee 2001) (local adaptation of interpolation to the properties of the relaxation procedure).

The improvement of individual components of a multigrid method applied to solving nonlinear equations (e.g., the Euler or Navier–Stokes equations) weakly affects the convergence rate (especially when high-order difference schemes, which are not compact, are used). To improve the convergence of a multigrid procedure, the Jacobian must be discretized more accurately (Mavriplis 2002) (the Jacobian is typically discretized using low-order schemes, which results in the degradation of convergence).

5.5 *Numerical Examples*

Consider the inviscid and viscous flow around the NACA0012 airfoil at $M_\infty = 0.5$ and at different attack angles ($\alpha = 0°$ and $\alpha = 1.25°$). The Euler and Navier–Stokes equations are discretized using structured and unstructured meshes. The multigrid procedure is implemented within the V-cycle ($\mu_1 = 2$ and $\mu_2 = 1$). The three-stage (RK3) and five-stage (RK5) Runge–Kutta method is used for smoothing.

5.5.1 Geometric Method

Computations for $\alpha = 1.25°$ are performed on a structured mesh (version 1) containing 20,800 nodes (320×64 mesh of type O) and on an unstructured mesh consisting of triangular cells (version 2) containing 5766 nodes. In version 1, the outflow boundary of the computation domain is at the distance of $20L$ from the trailing edge of the airfoil. In the case of version 2, it is at the distance of $30L$ from the trailing edge, where L is the airfoil chord. The finest mesh cells satisfy the conditions $(\Delta y/\Delta x)_{\min} = 0.45$, $(\Delta y/\Delta x)_{\max} = 20$, and $(\Delta y/L) = 8 \times 10^{-4}$. While the free stream conditions are set on the inflow boundaries, those of the outflow are set on the outlet boundaries.

Five mesh levels are used in CFD computations. The meshes are constructed by collapsing mesh edges in all directions. The meshes of levels 1, 2, and 3 constructed using the edge collapsing algorithm are shown in Fig. 27 (version 1) and Fig. 28 (version 2). The number of nodes in the meshes of each level is shown in Table 1

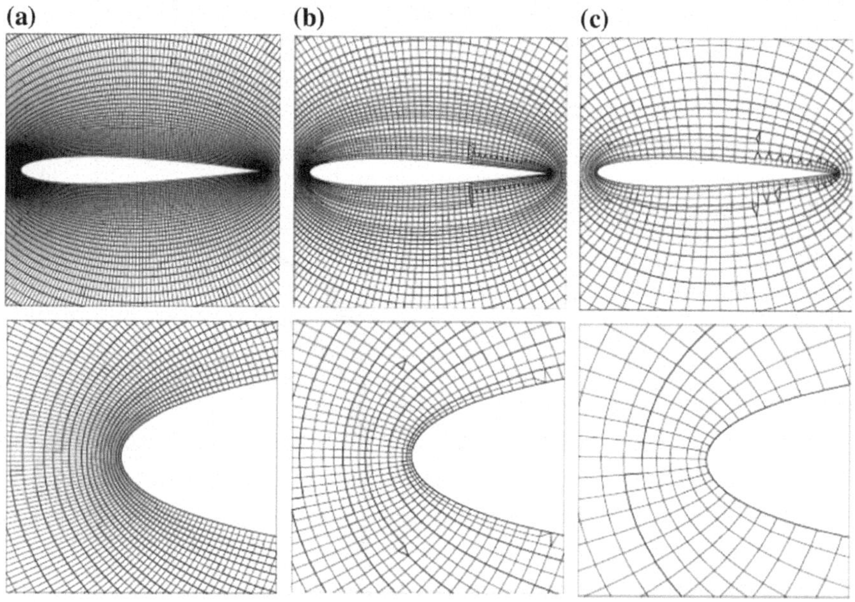

Fig. 27 Sequence of structured meshes (version 1)

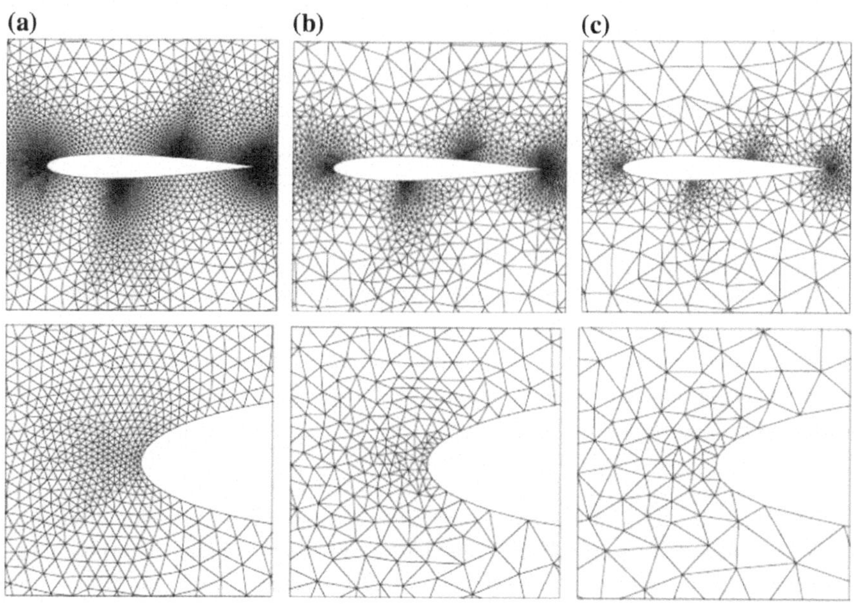

Fig. 28 Sequence of unstructured meshes (version 2)

Table 1 The number of mesh nodes at each level

Variant	Level 1	Level 2	Level 3	Level 4	Level 5
1	20,800	9496	3767	1825	603
2	5766	2180	893	403	214

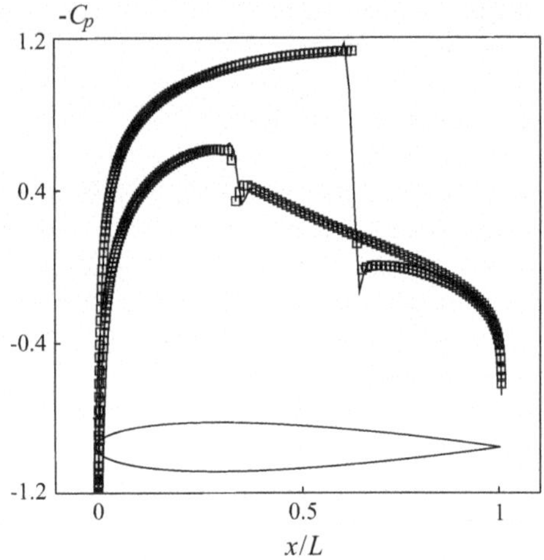

Fig. 29 Comparison of the distributions of pressure coefficient over the airfoil surface on structured (solid line) and unstructured (symbols □) meshes

Table 2 Convergence of the iterative process

Variant	Convergence $10^0 \to 10^{-4}$		Convergence $10^0 \to 10^{-10}$	
	Number of cycles	Time (s)	Number of cycles	Time (s)
1	180	1072	485	2560
2	267	269	876	833

(for versions 1 and 2). As the mesh is coarsened, the leading edge of the airfoil undergoes certain changes. It becomes sharper, which introduces errors in the coarse mesh solution but does not affect the solution obtained by the multigrid technique. The procedure used to construct the nested meshes preserves the mesh topology inside the computation domain. In version 2, the maximum ratio of the cell edges is 1 for the first level mesh, 0.37 for the second level mesh, 0.44 for the third level mesh, 0.57 for the fourth level mesh, and 0.71 for the fifth level mesh.

The distributions of the pressure coefficient on the airfoil surface are shown in Fig. 29. They look almost the same both on the structured and unstructured meshes. The solution has a big jump on the upper surface of the airfoil (at $x/L \sim 0.68$) and a weak jump on its lower surface (at $x/L \sim 0.32$). For the given parameters of the problem, the integration of pressure over the airfoil surface yields the coefficients of drag and lift force ($C_x = 0.0227$ and $C_y = 0.3527$) that are in good agreement with the data presented in AGARD Report (1986).

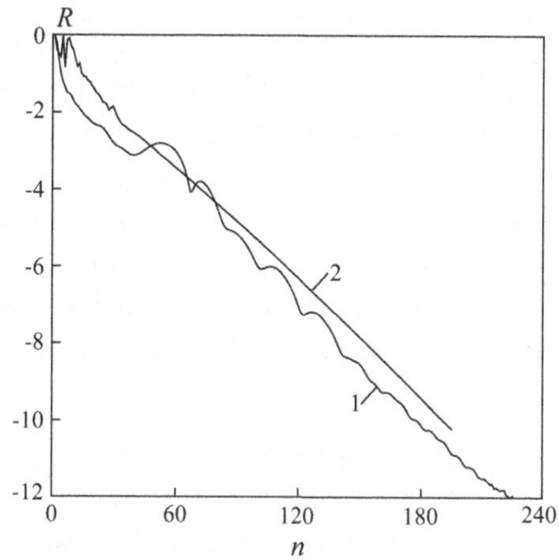

Fig. 30 Variation of residual depending on the number of multigrid cycles on structured (line 1) and unstructured (line 2) meshes

The data for the convergence of the iterative process in the initial stage of the residual variation (from 10^0 to 10^{-4}) and in the entire range of its variation are presented (from 10^0 to 10^{-10}) in Table 2 (the residual norm is normalized to the norm of the initial residual). The number of iterations for the preliminary and final smoothing is assumed to be equal to one. On the coarsest mesh, five iterations are made.

The history of convergence on the structured and unstructured meshes is illustrated in Fig. 30. On the structured mesh (version 1), the method converges in 485 multigrid cycles, and on the unstructured mesh (version 2), it converges in 876 multigrid cycles.

The solution of the viscous problem is performed at $Re = 500$ (the Reynolds number is calculated based on the airfoil chord). Calculations are based on Reynolds-averaged Navier–Stokes equations and equations of Spalart–Allmaras turbulence model (Volkov 2008). The simulation gives the distribution of pressure over the airfoil surface that agrees with the data in AGARD Report (1986) and only weakly differs from the solution shown in Fig. 29. The drag and lift force coefficients are $C_x = 0.0225$ and $C_y = 0.3536$ (the relative difference in the solutions to the inviscid and viscous problems is 0.88% for the drag coefficient and 0.26% for

Table 3 The convergence factor in the solution of the Euler equations

Mesh	Number of levels	C_{Gv}	C_{Gc}	$\rho(\alpha = 0°)$		$\rho(\alpha = 1.25°)$	
				RK3	RK5	RK3	RK5
2607	2	1.58	1.34	0.24	0.18	0.28	0.18
5258	3	1.64	1.36	0.34	0.24	0.25	0.17
10,273	4	1.67	1.36	0.25	0.24	0.24	0.17
20,621	5	1.64	1.36	0.34	0.21	0.23	0.21

Table 4 The convergence factor in the solution of the Navier–Stokes equations

Mesh	Number of levels	C_{Gv}	C_{Gc}	$\rho(\alpha = 0°)$	$\rho(\alpha = 1.25°)$
8872	2	1.64	1.37	0.37	0.21
18,416	3	1.88	1.44	0.37	0.57
36,388	4	1.67	1.37	0.35	0.45

the lifting force coefficient). To achieve the prescribed level of relative residual ($R \sim 10^{-5}$), 158 multigrid cycles are needed.

5.5.2 Algebraic Method

The C/F splitting is constructed using the standard coarsening method. For interpolation, the standard interpolation method is used, and smoothing is performed by the Gauss–Seidel method.

The convergence factor of the multigrid procedure is presented in Table 3. The quality of mesh coarsening is characterized by such parameters as the total number of mesh cells and the number of mesh nodes in the meshes of different resolution (the parameters C_{Gc} and C_{Gv}). The coarsest mesh contains less than 500 unknowns. The dependence of the convergence rate on the mesh resolution is relatively weak.

The simulation of the viscous airfoil flow is performed for $Re = 500$. The mesh is refined in the vicinity of the airfoil surface. The maximum ratio of the mesh cell sizes is 500. The scheme RK5 is used as the smoothing method. The convergence factor of the multigrid procedure is presented in Table 4. The convergence rate at the attack angle $\alpha = 0°$ is higher than that of $\alpha = 1.25°$.

The use of the algebraic multigrid gives the convergence factor of 0.55 for the Euler equations and 0.81 for the Navier–Stokes equations. For the geometric multigrid technique, the asymptotic convergence factor is 0.8 for the Euler equations and 0.91 for the Navier–Stokes equations.

In the majority of implementations, the parameter $\theta = 0.25$ is used, and the number of mesh levels is assigned the maximum possible value (Ruge and Stuben 1987). In practice, the coefficient θ is varied in a wide range (from 0.2 through 0.4). The value $\theta = 0.25$ corresponds to the interpolation by four points on the fine mesh (in the geometric method, this value corresponds to doubling the mesh step when going from a finer to a coarser mesh). The variation of θ in the range from 0.2 through 0.4 gives the maximum increase in the computation time by 7.5% on the structured mesh. For the unstructured meshes, the optimal value is 0.0625. The choice of the optimal number of internal iterations reduces the computation time by 10% depending on the boundary conditions.

6 Conclusion

A numerical method was developed for computing steady-state inviscid and viscous compressible flows in a wide range of Mach numbers. Preconditioning is switched on depending on the local Mach number or the local pressure field. The numerical results obtained in the benchmark problems suggest that the numerical method developed has a sufficient accuracy for resolving characteristic features of incompressible and compressed flows. At low Mach numbers, the CPU time required for solving the preconditioned equations is 15% longer (due to an increase in the number of arithmetic operations) than in the case of the original equations.

The dual time-stepping method is used to compute free convective flows. The proposed numerical method has sufficient accuracy to solve characteristic features of free convective flows. Preconditioning eliminates the dependence of the convergence rate on the Mach number. At low Mach numbers, the solution of preconditioned equations increases the computation time 10–15% (due to the growing number of arithmetic operations) compared with the case when no preconditioning is used, and the original equations are solved.

The results obtained show that the geometric multigrid is more efficient than the algebraic multigrid, even though they require more effort for the modification of the software code, including the modification of the data structure (arrays for each mesh level must be organized and the procedure for the construction of the sequence of meshes must be implemented). The geometric multigrid has a higher convergence rate and does not strongly depend on the method parameters. The algebraic multigrid efficiency considerably depends on the input parameters. The geometric method takes about 20% of the computation time needed to solve the same problem by the algebraic method (the time needed to construct the meshes is not taken into account). The algebraic method is sensitive to the mesh variation and resolution on the coarsest level. The proper choice of the algebraic multigrid method parameters can reduce the total computation time by 15–40%.

Nomenclature

Latin Symbols

c	Speed of sound
e	Specific total energy
f	Function
m	Mass
n	Number of iterations
p	Pressure
q	Heat flux
r	Radius
t	Time
v_x, v_y, v_z	Velocity components
v	Velocity vector
x, y, z	Cartesian coordinates

C	Constant
C_f	Friction coefficient
C_p	Pressure coefficient
F	Flux
H	Source term
L	Total pressure losses
M	Mach number
Nu	Nusselt number
Pr	Prandtl number
Q	Conservative variables
R	Residual
Ra	Rayleigh number
Re	Reynolds number
T	Temperature
U	Symmetrized variables
V	Primitive variables
W	Characteristic variables

Greek Symbols

γ	Ratio of specific heat capacities
λ	Thermal conductivity
μ	Dynamic viscosity
ν	Kinematic viscosity
ρ	Density
ω	Rotation speed

Subscripts

i, j, k	Tensor indices
n	Normal
w	Wall
τ	Tangential
0	Total
$*$	Reference
∞	Free stream

Superscripts

n	Time layer

Abbreviations

1D	One-dimensional
2D	Two-dimensional
3D	Three-dimensional
CDS	Central difference scheme
CFD	Computational fluid dynamics
CFL	Courant–Friedrichs–Lewy
MG	Multigrid
MUSCL	Monotonic upwind scheme for conservative laws
RK	Runge–Kutta

Acknowledgements This work was financially supported by the Ministry of Education and Science of the Russian Federation (agreement No 14.578.21.0203, a unique identifier of applied scientific research RFMEFI57816X0203).

Appendix: Physical and Conservative Variables

In integral form, the system of gas dynamics equations describing unsteady viscous compressible gas flows has the form

$$\frac{\partial}{\partial t}\int\limits_{V_i} U\,\mathrm{d}V + \int\limits_{\partial V_i} (F+G)\,\mathrm{d}S = 0.$$

In physical variables, it becomes

$$\frac{\partial U}{\partial Q}\frac{\partial}{\partial t}\int\limits_{V_i} Q\,\mathrm{d}V + \int\limits_{\partial V_i} (F+G)\mathrm{d}S = 0.$$

The vectors of physical and conservative variables are given by

$$Q = \begin{pmatrix} p \\ v_x \\ v_y \\ v_z \\ T \end{pmatrix}, \quad U = \begin{pmatrix} \rho \\ \rho v_x \\ \rho v_y \\ \rho v_z \\ \rho e \end{pmatrix}.$$

The transition between the conservative and physical variables is based on the matrix

$$\frac{\partial U}{\partial Q} = \begin{pmatrix} \rho_{\mathrm{p}} & 0 & 0 & 0 & \rho_{\mathrm{T}} \\ \rho_{\mathrm{p}} v_x & \rho & 0 & 0 & \rho_{\mathrm{T}} v_x \\ \rho_{\mathrm{p}} v_y & 0 & \rho & 0 & \rho_{\mathrm{T}} v_y \\ \rho_{\mathrm{p}} v_z & 0 & 0 & \rho & \rho_{\mathrm{T}} v_z \\ \rho_{\mathrm{p}} H - 1 & \rho v_x & \rho v_y & \rho v_z & \rho_{\mathrm{T}} H + \rho c_{\mathrm{p}} \end{pmatrix}.$$

Preconditioning Matrix

Multiplication of the equation written in physical variables by a matrix K yields the following equation in nonconservative form:

$$\left(K \frac{\partial U}{\partial Q}\right) \frac{\partial}{\partial t} \int_{V_i} Q \, \mathrm{d}V + K \int_{\partial V_i} F \, \mathrm{d}S = 0.$$

The matrix K is chosen so as to simplify the product

$$K \frac{\partial U}{\partial Q} = \begin{pmatrix} \rho_{\mathrm{p}} & 0 & 0 & 0 & \rho_{\mathrm{T}} \\ 0 & \rho & 0 & 0 & 0 \\ 0 & 0 & \rho & 0 & 0 \\ 0 & 0 & 0 & \rho & 0 \\ -1 & 0 & 0 & 0 & \rho c_{\mathrm{p}} \end{pmatrix}.$$

It has the form

$$K = \begin{pmatrix} 1 & 0 & 0 & 0 & 0 \\ -v_x & 1 & 0 & 0 & 0 \\ -v_y & 0 & 1 & 0 & 0 \\ -v_z & 0 & 0 & 1 & 0 \\ -(H - |v|^2) & -v_x & -v_y & -v_z & 1 \end{pmatrix}.$$

The inverse matrix of K is written as

$$K^{-1} = \{R_1, R_2, R_3, R_4, R_5\}.$$

The columns of K^{-1} have the form

$$R_1 = \begin{pmatrix} U_r^2[b(H-|v|^2)+1]/a \\ -v_x/e \\ -v_y/e \\ -v_z/e \\ \left\{(m+1)U_r^2+(H-|v|^2)[(m+1)bU_r^2-a]\right\}/(c_{\mathrm{p}}ae) \end{pmatrix},$$

$$R_2 = \begin{pmatrix} bU_r^2 v_x/a \\ 1/e \\ 0 \\ 0 \\ \left[(m+1)bU_r^2-a\right]v_x/(c_{\mathrm{p}}ae) \end{pmatrix},$$

$$R_3 = \begin{pmatrix} bU_r^2 v_y/a \\ 0 \\ 1/e \\ 0 \\ \left[(m+1)bU_r^2-a\right]v_y/(c_{\mathrm{p}}ae) \end{pmatrix},$$

$$R_4 = \begin{pmatrix} bU_r^2 v_z/a \\ 0 \\ 0 \\ 1/e \\ \left[(m+1)bU_r^2-a\right]v_z/(c_{\mathrm{p}}ae) \end{pmatrix},$$

$$R_5 = \begin{pmatrix} -bU_r^2/a \\ 0 \\ 0 \\ 0 \\ -\left[(m+1)bU_r^2-a\right]/(c_{\mathrm{p}}ae) \end{pmatrix}.$$

Here, $a = 1 + m\beta U_r^2$, $b = \rho_{\mathrm{T}}/(\rho c_{\mathrm{p}})$, $e = (m+1)\rho$, and v is the velocity vector.

The modified matrix $\widetilde{\Gamma} = K(\partial U/\partial Q)$ for the system of nonconservative equations is given by

$$\widetilde{\Gamma} = \begin{pmatrix} \Theta & 0 & 0 & 0 & \rho_{\mathrm{T}} \\ 0 & \rho & 0 & 0 & 0 \\ 0 & 0 & \rho & 0 & 0 \\ 0 & 0 & 0 & \rho & 0 \\ -1 & 0 & 0 & 0 & \rho c_{\mathrm{p}} \end{pmatrix}.$$

Multiplication of the equation by the matrix K^{-1}, the conservative preconditioned equation is

$$\Gamma \frac{\partial}{\partial t} \int\limits_{V_i} Q \,\mathrm{d}V + \int\limits_{\partial V_i} F \,\mathrm{d}S = 0,$$

where $\Gamma = K^{-1}\widetilde{\Gamma}$. The inverse of the matrix Γ has the form

$$\Gamma^{-1} = \begin{pmatrix} ab + U_r^2 & av_x & av_y & av_z & -a \\ -v_x/\rho & 1/\rho & 0 & 0 & 0 \\ -v_y/\rho & 0 & 1/\rho & 0 & 0 \\ -v_z/\rho & 0 & 0 & 1/\rho & 0 \\ [b(a-1) + U_r^2]/(\rho c_{\mathrm{p}}) & ev_x & ev_y & ev_z & -e \end{pmatrix},$$

where $a = \rho_{\mathrm{T}} U_r^2/(\rho c_{\mathrm{p}})$, $b = H - |v|^2$, and $e = (a-1)/(\rho c_{\mathrm{p}})$.

The eigenvalues of the Jacobian of the preconditioned system are found by solving the characteristic equation

$$\left|\Gamma^{-1}A - \lambda I\right| = |A - \lambda\Gamma| = 0,$$

where $A = \partial F_x/\partial Q$. The flux Jacobian in physical variables has the form

$$A = \begin{pmatrix} \rho_{\mathrm{p}} v_x & \rho & 0 & 0 & \rho_{\mathrm{T}} v_x \\ \rho_{\mathrm{p}} v_x^2 + 1 & 2\rho v_x & 0 & 0 & \rho_{\mathrm{T}} v_x^2 \\ \rho_{\mathrm{p}} v_x v_y & \rho v_y & \rho v_x & 0 & \rho_{\mathrm{T}} v_x v_y \\ \rho_{\mathrm{p}} v_x v_z & \rho v_z & 0 & \rho v_x & \rho_{\mathrm{T}} v_x v_z \\ \rho_{\mathrm{p}} H v_x & \rho(H + v_x^2) & \rho v_x v_y & \rho v_x v_z & (\rho_{\mathrm{T}} H + \rho c_{\mathrm{p}}) v_x \end{pmatrix}.$$

Rearranging the determinant gives

$$\begin{vmatrix} \rho_p v_x - \lambda\Theta & \rho & 0 & 0 & \rho_T(v_x - \lambda) \\ 1 & \rho(v_x - \lambda) & 0 & 0 & 0 \\ 1 & 0 & \rho(v_x - \lambda) & 0 & 0 \\ 1 & 0 & 0 & \rho(v_x - \lambda) & 0 \\ \lambda & \rho v_x(v_x - \lambda) & \rho v_y(v_x - \lambda) & \rho v_z(v_x - \lambda) & \rho c_p(v_x - \lambda) \end{vmatrix} = 0.$$

The characteristic equation becomes

$$(v_x - \lambda)^3 [\rho c_p - v_x^2(\rho_T + \rho\rho_p c_p) + \lambda v_x(2\rho_T + \rho\rho_p c_p + \Theta\rho c_p) - \lambda^2(\rho_T + \Theta\rho c_p)] \\ = 0.$$

Here, $\lambda_{1,2,3} = v_x$. The roots $\lambda_{4,5}$ are found by solving the quadratic equation

$$\lambda^2 - v_x(1 + \beta U_r^2)\lambda - U_r^2(1 - \beta v_x^2) = 0,$$

where $\beta = \rho_p + \rho_T/(\rho c_p)$. For the ideal gas model, $\beta = 1/(\gamma RT) = 1/c^2$. Solving the quadratic equation yields

$$\lambda_{4,5} = v_x' \mp c',$$

where

$$v_x' = v_x(1 - \alpha), \quad c' = \left(\alpha^2 v_x^2 + U_r^2\right)^{1/2}.$$

The parameter α is given by the relation

$$\alpha = \frac{1}{2}\left(1 - \beta U_r^2\right).$$

Eigenvalues and Eigenvectors

The right eigenvectors of the matrix A_Γ are determined by the relation $A_\Gamma r_i = \lambda_i r_i$, which gives

$$(A - \lambda_i \Gamma) r_i = 0.$$

Substitution and simple transformations give

$$\begin{pmatrix} \rho_p v_x - \lambda\Theta & \rho & 0 & 0 & \rho_T(v_x-\lambda) \\ 1 & \rho(v_x-\lambda) & 0 & 0 & 0 \\ 0 & 0 & \rho(v_x-\lambda) & 0 & 0 \\ 0 & 0 & 0 & \rho(v_x-\lambda) & 0 \\ \lambda & \rho v_x(v_x-\lambda) & \rho v_y(v_x-\lambda) & \rho v_z(v_x-\lambda) & \rho c_p(v_x-\lambda) \end{pmatrix} \begin{pmatrix} r_1 \\ r_2 \\ r_3 \\ r_4 \\ r_5 \end{pmatrix}$$
$$= 0.$$

At $\lambda = v_x$, the equation becomes

$$\begin{pmatrix} \rho_p v_x - v_x\Theta & \rho & 0 & 0 & 0 \\ 1 & 0 & 0 & 0 & 0 \\ 0 & 0 & 0 & 0 & 0 \\ 0 & 0 & 0 & 0 & 0 \\ \lambda & 0 & 0 & 0 & 0 \end{pmatrix} \begin{pmatrix} r_1 \\ r_2 \\ r_3 \\ r_4 \\ r_5 \end{pmatrix} = 0.$$

Here, $r_1 = 0$, $r_2 = 0$, while the components r_3, r_4, and r_5 are chosen to be linearly independent

$$r_1 = \begin{pmatrix} 0 \\ 0 \\ 0 \\ 0 \\ 1 \end{pmatrix} 0, \quad r_2 = \begin{pmatrix} 0 \\ 0 \\ 0 \\ 1 \\ 0 \end{pmatrix}, \quad r_3 = \begin{pmatrix} 0 \\ 0 \\ 1 \\ 0 \\ 0 \end{pmatrix}.$$

For the components of the right eigenvectors r_4 and r_5 with eigenvalues $\lambda_4 = v'_x - c'$, $\lambda_5 = v'_x + c'$

$$r_2 = \frac{r_1}{\rho(\lambda - v_x)}, \quad r_3 = 0, \quad r_4 = 0, \quad r_5 = \frac{r_1}{\rho c_p}.$$

Since $\lambda = u' - c'$, then $\lambda - u = -(\alpha u + c')$ and $r_1 = \alpha u + c'$. The eigenvector corresponding to the eigenvalue λ_4 is given by

$$r_4 = \begin{pmatrix} \alpha u + c' \\ -1/\rho \\ 0 \\ 0 \\ (\alpha u + c')/(\rho c_p) \end{pmatrix}.$$

For the eigenvalue $\lambda_5 = u' + c'$, $\lambda - u = -(\alpha u - c')$. Setting $r_1 = \alpha u - c'$ yields

$$r_5 = \begin{pmatrix} \alpha u - c' \\ -1/\rho \\ 0 \\ 0 \\ (\alpha u - c')/(\rho c_p) \end{pmatrix}.$$

The right eigenvector matrix is written as

$$M_\Gamma = \begin{pmatrix} 0 & 0 & 0 & (\alpha u + c') & (\alpha u - c') \\ 0 & 0 & 0 & -1/\rho & -1/\rho \\ 0 & 0 & 1 & 0 & 0 \\ 0 & 1 & 0 & 0 & 0 \\ 1 & 0 & 0 & (\alpha u + c')/(\rho c_p) & (\alpha u - c')/(\rho c_p) \end{pmatrix}.$$

The left eigenvector matrix (the inverse of M_Γ) has the form

$$M_\Gamma^{-1} = \begin{pmatrix} -1/(\rho c_p) & 0 & 0 & 0 & 1 \\ 0 & 0 & 0 & 1 & 0 \\ 0 & 0 & 1 & 0 & 0 \\ 1/(2c') & \rho(\alpha u - c')/(2c') & 0 & 0 & 0 \\ -1/(2c') & -\rho(\alpha u + c')/(2c') & 0 & 0 & 0 \end{pmatrix}.$$

The eigenvalue matrix Λ is given by

$$\Lambda = \begin{pmatrix} v_x & 0 & 0 & 0 & 0 \\ 0 & v_x & 0 & 0 & 0 \\ 0 & 0 & v_x & 0 & 0 \\ 0 & 0 & 0 & u - (\alpha u + c') & 0 \\ 0 & 0 & 0 & 0 & u - (\alpha u - c') \end{pmatrix}.$$

The matrix $A_\Gamma = \Gamma^{-1} A$ is written as

$$A_\Gamma = \begin{pmatrix} v_x \beta U_r^2 & \rho U_r^2 & 0 & 0 & 0 \\ 1/\rho & v_x & 0 & 0 & 0 \\ 0 & 0 & v_x & 0 & 0 \\ 0 & 0 & 0 & v_x & 0 \\ -2\alpha v_x/(\rho c_p) & U_r^2/c_p & 0 & 0 & v_x \end{pmatrix}.$$

Runge–Kutta Scheme

In the general case, the explicit Runge–Kutta scheme is written as

$$Q_i^{(0)} = Q_i^n;$$
$$Q_i^{(k)} = Q_i^{(0)} - \alpha_k \Delta t \Gamma R_i[Q^{(k-1)}];$$
$$Q_i^{n+1} = Q_i^{(m)}.$$

Here, $\alpha_k = 1/(m-k+1)$, $k = 1, 2, \ldots, m$. The indexes n and m denote the time layer and the order of the Runge–Kutta scheme, respectively. R denotes the residual due to the discretization of inviscid and viscous fluxes.

References

AGARD Fluid Dynamics Panel (1986) Test cases for inviscid flow field methods. AGARD advisory reports AR-211

Bijl H, Carpenter MH, Vatsa VN, Kennedy CA (2002) Implicit time integration schemes for the unsteady compressible Navier–Stokes equations: laminar flow. J Comput Phys 179(1):313–329

Brandt A (1977) Multi-level adaptive solutions to boundary value problems. Math Comput 31 (138):46–50

Brandt A (1984) Guide to multigrid development. Lect Notes Math 960:220–312

Brannick J, Brezina M, Falgout R, Manteuffel T, McCormick S, Ruge J, Sheehan B, Xu J, Zikatanov L (2006) Extending the applicability of multigrid methods. J Phys Conf Ser 46: 443–452

Briley WR, McDonald H (1977) Solution of the multidimensional compressible Navier–Stokes equations by a generalized implicit method. J Comput Phys 24(4):372–397

Buelow PEO, Venkateswaran S, Merkle CL (1984) Effect of grid aspect ratio on convergence. AIAA J 32(12):2402–2408

Char M, Hsu YH (1998) Comparative analysis of linear and nonlinear low-Reynolds-number eddy viscosity models to turbulent natural convection in horizontal cylindrical annuli. Numer Heat Transf 33:191–206

Choi D, Merkle CL (1985) Application of time-iterative schemes to incompressible flow. AIAA J 23(10):1518–1524

Choi Y-H, Merkle CL (1993) The application of preconditioning in viscous flows. J Comput Phys 105(2):207–223

Chorin AJ (1967) A numerical method for solving incompressible viscous flow problems. J Comput Phys 2(1):12–26

Chorin AJ (1968) Numerical solution of the Navier-Stokes equations. Math Comput 22(104): 745–762

Cleary AJ, Falgout RD, Henson VE, Jones JE, Manteu TA, McCormick SF, Miranda GN, Ruge JW (2000) Robustness and scalability of algebraic multigrid. SIAM J Sci Stat Comput 21 (6):1886–1908

Colin Y, Deniau H, Boussuge J-F (2011) A robust low speed preconditioning formulation: application to air intake flows. Comput Fluids 47(1):1–15

Crumpton PI, Moinier P, Giles MB (1997) An unstructured algorithm for high Reynolds number flows on highly stretched grids. In: Numerical methods in laminar and turbulent flows. Pineridge Press, pp 561–572

Darmofal DL, Schmid PJ (1996) The importance of eigenvectors for local preconditioners of the Euler equations. J Comput Phys 127(2):728–756

Darmofal DL, Moinier P, Giles MB (2000) Eigenmode analysis of boundary conditions for the one-dimensional preconditioned Euler equations. J Comput Phys 160(1):369–384

Desai CP, Vafai K (1994) An investigation and comparative analysis of two and three-dimensional turbulent natural convection in a horizontal annulus. Int J Heat Mass Transf 37(16):2475–2504

Eidelman S, Colella P, Shreeve RP (1984) Application of the Godunov method and its second-order extension to cascade flow modeling. AIAA J 22(11):1609–1615

Emelyanov VN, Karpenko AG, Volkov KN (2015) Development of advanced computational fluid dynamics tools and their application to simulation of internal turbulent flows. Prog Flight Phys 7:247–268

Fedorenko RP (1961) A relaxation method for solving elliptic difference equations. Comput Math Math Phys 1(4):922–927

Fu W-S, Li C-G, Huang C-P, Huang J-C (2009) An investigation of a high temperature difference natural convection in a finite length channel without Bossinesq assumption. Int J Heat Mass Transf 52(11–12):2571–2580

Gray DD, Giorgini A (1976) The validity of the Boussinesq approximation for liquids and gases. Int J Heat Mass Transf 19(5):545–551

Guillard H, Viozat C (1999) On the behaviour of the upwind schemes in the low Mach number limit. Comput Fluids 28(1):63–86

Harten A (1983) High resolution schemes for hyperbolic conservation laws. J Comput Phys 49 (3):357–393

Housman J, Kiris C, Hafez M (2009) Preconditioned methods for simulations of low speed compressible flows. Comput Fluids 38(7):1411–1423

Jameson A (1983) Solution of the Euler equations by a multigrid method. Appl Math Comput 13 (3–4):327–356

Jameson A (1991) Time dependent calculations using multigrid, with application to unsteady flows past airfoils and wings. AIAA paper 91-1596

Jameson A (1995) Analysis and design of numerical schemes for gas dynamics: artificial diffusion, upwind biasing, limiters and their effect on accuracy and multigrid convergence. Int J Comput Fluid Dyn 4(3–4):171–218

Jameson A (2009) An assessment of dual-time stepping, time spectral and artificial compressibility based numerical algorithms for unsteady flow with applications to flapping wings. AIAA paper 2009-4273

Kuehn TH, Goldstein RJ (1976) An experimental and theoretical study of natural convection in the annulus between horizontal concentric cylinders. J Fluid Mech 74:695–719

Kuehn TH, Goldstein RJ (1978) An experimental study of natural convection heat transfer concentric and eccentric horizontal cylindrical annuli. J Heat Transf 100:635–640

Kwak D, Kiris C, Housman J (2011) Implicit methods for viscous incompressible flows. Comput Fluids 41(1):51–64

Lee S-H (2007) Cancellation problem of preconditioning method at low Mach numbers. J Comput Phys 225(2):1199–1210

Luo H, Baum JD, Löhner R (2001) An accurate, fast, matrix-free implicit method for computing unsteady flows on unstructured grids. Comput Fluids 30(2):137–159

MacCormack RW (1985) Current status of numerical solutions of the Navier–Stokes equations. AIAA paper 85-0032

Mavriplis DJ (2002) An assessment of linear versus nonlinear multigrid methods for unstructured mesh solvers. J Comput Phys 175(1):302–325

Merkle CL, Choi Y-H (1988) Computation of low-speed compressible flows with time marching procedures. Int J Numer Methods Eng 25(2):293–311

Moinier P, Giles MB (2001) Compressible Navier-Stokes equations for low Mach number applications. In: Proceedings of the ECCOMAS computational fluid dynamics conference, 4–7 Sept 2001, Swansea, United Kingdom

Müller J-D, Giles MB (1998) Edge-based multigrid schemes for hybrid grids. Numer Methods Fluid Dyn 6:425–432
Patankar SV, Spalding DB (1972) A calculation procedure for heat, mass and momentum transfer in three-dimensional parabolic flows. Int J Heat Mass Transf 15(10):1787–1806
Pierce NA, Giles MB (1996) Preconditioning compressible flow calculations on stretched meshes. AIAA paper 96-0889
Pierce NA, Giles MB (1997) Preconditioned multigrid method for compressible flow calculations on stretched meshes. J Comput Phys 136(2):425–445
Pierce NA, Giles MB, Jameson A, Martinelli L (1997) Accelerating three-dimensional Navier–Stokes calculations. AIAA paper 97-1953
Rogers SE, Kwak D (1990) Upwind differencing scheme for the time-accurate incompressible Navier–Stokes equations. AIAA J 28(2):253–262
Ruge J, Stüben K (1987) Algebraic multigrid (AMG). Multigrid methods. Front Appl Math (SIAM, Philadelphia) 3:73–130
Rumsey CL, Sanetrik MD, Biedron RT, Melson ND, Parlette EB (1996) Efficiency and accuracy of time-accurate Navier–Stokes computations. Comput Fluids 25(2):217–236
Shuen J-S, Chen K-H, Choi Y (1993) A coupled implicit method for chemical non-equilibrium flows at all speeds. J Comput Phys 106(2):306–318
Stüben K (2001a) A review of algebraic multigrid. J Comput Appl Math 128(1–2):281–309
Stüben K (2001b) An introduction to algebraic multigrid. In: Multigrid. Academic Press, London, pp 413–532
Tang HS, Jones SC, Sotiropoulos F (2003) An overset grid method for 3D, unsteady, incompressible flows. J Comput Phys 191(2):567–600
Trottenberg U, Oosterlee C, Schüller A (2001) Multigrid. Academic Press, London
Turkel E (1987) Preconditioned methods for solving the incompressible and low speed compressible equations. J Comput Phys 72(1):277–298
Turkel E (1993) Review of preconditioning methods for fluid dynamics. Appl Numer Math 12(1–3):257–284
Turkel E (1997) Preconditioning-squared methods for multidimensional aerodynamics. AIAA paper 97-2025
Turkel E, Fiterman A, van Leer B (1993) Preconditioning and the limit to the incompressible flow equations. ICASE report 93-42
Turkel E, Vatsa V, Radespiel R (1996) Preconditioning methods for low-speed flows. AIAA paper 96-2640
Turkel E, Radespiel R, Kroll N (1997) Assessment of preconditioning methods for multidimensional aerodynamics. Comput Fluids 26(6):613–634
van Leer B, Lee W, Roe P (1991) Characteristic time-stepping or local preconditioning of the Euler equations. AIAA paper 91-1552
Venkateswaran S, Merkle CL (1995) Dual time stepping and preconditioning for unsteady computations. AIAA paper 95-0078
Venkateswaran S, Merkle CL (1999) Analysis of preconditioning methods for the Euler and Navier–Stokes equations. Von Karman Institute Lecture Series on CFD 1999-30
Volkov KN (2008) Unstructured-grid finite volume discretization of the Navier-Stokes equations based on high-resolution difference schemes. Comput Math Math Phys 48(7):1181–1202
Volkov KN (2009) Preconditioning of the Euler and Navier-Stokes equations in low-velocity flow simulation on unstructured grids. Comput Math Math Phys 49(10):1789–1803
Volkov KN (2010) Multigrid techniques as applied to gas dynamic simulation on unstructured meshes. Comput Math Math Phys 50(11):1837–1850
Volkov KN (2011) Simulation of turbulent flows in rotating disc cavity systems. Turbulence: theory, types and simulation. Nova Science, USA, pp 569–686
Volkov KN, Emelyanov VN, Teterina IV (2016) Geometrical and algebraic multigrid techniques for fluid dynamics problems on unstructured grids. Comput Math Math Phys 56(2):286–302
Volpe G (1993) Performance of compressible flow codes at low Mach numbers. AIAA J 31(1): 49–56

Weiss J, Smith W (1995) Preconditioning applied to variable and constant density flows. AIAA J 33(11):2050–2062
Weiss JM, Maruszewski JP, Smith WA (1999) Implicit solution of preconditioned Navier–Stokes equations using algebraic multigrid. AIAA J 37(1):29–36
Wesseling P (1992) An introduction to multigrid methods. Wiley, Chichester, p 1992
Wesseling P (2000) Principles of computational fluid dynamics. Springer, Berlin
Wesseling P, Oosterlee CW (2001) Geometric multigrid with applications to computational of fluid dynamics. J Comput Appl Math 128(1–2):311–334
Yang UM (2006) Parallel algebraic multigrid methods—high performance preconditioners. In: Numerical solution of partial differential equations on parallel computers, vol 51. Springer, Berlin, pp 209–236
Yoon S, Kwak D (1989) LU-SGS implicit algorithm for three-dimensional incompressible Navier–Stokes equations with source term. AIAA paper 89-1964
Zhang LP, Wang ZJ (2004) A block LU–SGS implicit dual time-stepping algorithm for hybrid dynamic meshes. Comput Fluids 33(7):891–916
Zhao Y (2004) Computation of complex turbulent flow using matrix-free implicit dual time-stepping scheme and LRN turbulence model on unstructured grids. Comput Fluids 33 (1):119–136
Zhao Y, Tan HH, Zhang BL (2002) A high-resolution characteristics-based implicit dual time-stepping of method for free surface flow simulation on unstructured grids. J Comput Phys 183(1):233–273

Numerical Simulation and Experimental Validation of the Role of Delta Wing Privileged Apex

Iddir Boumrar and Zied Driss

1 Introduction

The designers of the modern aircraft fighters have recently, given unprecedented interest for maneuverability, supersonic cruising, short takeoff, and the landing executions. In particular, supermaneuverability requires significant improvements of the aerodynamic characteristics of the wings at high angles of attack, which initially resulted in seeking geometries of optimal wings allowing and preserving a good dynamics of the turbulent flow at the suction face and eliminating each involved structure in the increase of the aerodynamic drag.

The behavior of the vortex occurring at the delta wing suction face at various incidences has been the subject of research since the beginning of the 50s. The research study developed by Parker (1976) is notable because it contains the complete list of the references and a discussion of the unstable flow for angles of attack going up to 20°.

Many investigators studied the air flow around delta wing models in wind tunnel at subsonic velocities with adapted volatile additives, which gave satisfaction. By employing air, we can reach high Reynolds numbers. For the delta wing model with sharp leading edges, we used Reynolds numbers varying from a few thousands until one or two millions. In a steady flow, when the delta wing incidence increases until a range between 30° and 40°, the exact value depends on the wing aspect ratio and other geometrical parameters; an instability of vortex known under the name of

I. Boumrar (✉)
Laboratory of Energétic Mechanics and Materials—LEMM,
University of Tizi-Ouzou, Tizi Ouzou, Algeria
e-mail: boumrari@hotmail.fr; boumrar_iddir@ummto.dz

Z. Driss
Laboratory of Electromechanical Systems (LASEM), University of Sfax,
Sfax, Tunisia
e-mail: Zied.Driss@enis.tn; Zied_Driss@yahoo.fr

Z. Driss et al. (eds.), *CFD Techniques and Thermo-Mechanics Applications*,
https://doi.org/10.1007/978-3-319-70945-1_7

"the bursting or vortex breakdown" appears at the wing trailing edge and progresses gradually ahead until the wing apex. On the delta wing model, this phenomenon was first observed by Lambourne and Bryer (1962). A study of Hawk et al. (1990) identified "a local angle of movements" around 45° from which the vortex bursting occurs. Ekaterinaris and Schiff (1990) showed that instability can be envisaged successfully (at $\alpha = 32°$ on a delta wing with a sweep angle $\varphi = 75°$) by an exact numerical resolution of the Navier–Stokes equations. These authors also mentioned two classes of vortex bursting known as the "bursting in bubble" and the "bursting in spiral". The behavior of the bursting is known to play a crucial role in the case of unstable movements. The first "generations" of planes, with delta wings, in the United States: the F-102, B-58, and the experimental bomber Xb-70 were not planned to fly at high angles of attack, they nevertheless drew all benefit from the increase of lift due to the apex vortex. Stability in pitching of the supersonic conveyor Concorde incited the first investigations on instabilities. These studies took place in the United Kingdom and were characterized by the work of Lambourne et al. (1964), which studied the consequences of the fast changes of the incidence angle i, like the oscillating data brought back later in detail by Woodgate and Halliday (1971). Laidlaw and Halfman (1956) have measured the pressures on oscillating models of delta wings with sweep angles $\varphi = 60°$ and 75°, but the amplitudes were small, and then the effect of the vortex was negligible; their objective was to evaluate the linear theory for the wings. However, it has reached incidences close to 90°, HARV used strongly beveled leading edges, and fighters in their preliminary design stage relied on the vortex mode coming from their leading edges (Cummings et al. 1990).

The theories published by Lowson (1963) and Randall (1966) were compared with experimental data. Lowson's analysis was particularly judged to be a rigorous attempt to adapt the work of Brown and Michael (1955) for oscillating movements. Nevertheless, these authors agree with the conclusions of Parker (1976) admitting that the errors between the experimental and theoretical results are significant adding that the comparisons have only a partial success.

It is not easy to quote all the contributions of those who have explored, through photographic and video techniques, the time dependent of the vortex movements. The principal contributions are given below, as simple lists. The order is not significant and it should be noted that some of the quoted articles also included data of forces or aerodynamic pressures. For example, Wang et al. (2003) have undertaken experiments in a water tunnel with 0.4 m of width, 0.4 m of depth, and 6.0 m of length. The experimental model is a delta wing of sweep angle $\varphi = 65°$; the wing leading edges are beveled. In order to facilitate the experimental observations, the suction face of the delta wing was also divided into five parts along the line of the chord, with radial lines resulting from the apex of the wing with a spacing of 10°. The jet is located at the center of the trailing edge with a rectangular exit nozzle of 2.4 cm × 0.3 cm. The variations of the jet orientation have been achieved by installing various jet openings in the trailing edge of the model. The experiments were considered at Reynolds numbers based on the chord of the wing of about 9.54×10^3. A dye was injected through 1-mm-diameter openings placed at the vicinity of the model apex to visualize the vortex bursting. An apparatus was used to record the flow models; the experiments were undertaken in a range of angles of attack $i = 25°–40°$ with a step of 5°. The exact position of vortex bursting is located at approximately 3–5 mm at the apex. In these experiments, the jet velocity changed from 0 to 10 m/s, and the jet angle from 0° to 60° toward the left side of the model. The motionless photographs show the hysteresis of the wake profiles and the vortex positions, with discussions of the flow details. Lang (2004) designed hypersonic vehicles using the delta wings configuration; ELAC was conceived as a system of transport for future orbital missions. Its first stage is carrying a body, and its second stage is to remain at an orbital altitude of approximately 30 km at $M = 7$. Initially, a combination of the oil flow and a vapor screen was applied to obtain visualizations of vortices at the top of the wing. The visualization of flow and PIV measurements are taken in the transonic wind tunnel with a 40 × 40-cm^2-sized test section. It is a wind tunnel with intermittent operation, which allows periods of approximately 3 s with Mach numbers of $M = 0.2–4$. The flow visualization at a relative length of chord $x/l = 30\%$ was carried out on a 1:100 scale model. The flow qualitative research at control faces of the ELAC and PIV measurements was carried out on a 1:240 scale model of a delta wing with sweep angle $\varphi = 75°$, with rounded leading edges to reduce the heat flow at the conditions of hypersonic flight. The Reynolds numbers are about $Re = 3.7 \times 10^6$, with Mach numbers $M = 2.0$ to $M = 2.5$. The experimental techniques applied were also used for the numerical simulation results validation. Many values on the position of the vortex core and its bursting are given.

Brodetsky et al. (2001) developed experiments in a supersonic wind tunnel where a maximum Mach number of 6 was reached. The tests were carried out on three delta wing models with respective sweep angles $\varphi = 68°$, 73°, and 78° and chords 383, 439, and 526 mm, at Mach numbers $M = 2–4$ and the angles of attack varying from $\alpha = 0°$ to 22°. The testing methods included the measurement of the static pressure on the model face and the flow visualization by the laser sheet technique. Photographs of the laser sheet and oil flow show the evolution of the swirling structure at a fixed angle of attack α. Also, the shock wave positions, size, and the position of primary and secondary vortex were obtained. Some new flow modes around delta wing were identified, and several visualizations were examined. Tests were carried out by Konrath et al. (2008) in a transonic wind tunnel; the test section has a size of 1 m × 1 m and communicates with a room, in which the total pressure can be placed in one range from 30,000 to 150,000 Pa. The test section was perforated, to give access for small indicators installed on top and on the lower wall behind which cameras, and PSP light sources were placed. The delta wing of sweep angle $\varphi = 65°$ provided by NASA was equipped with beveled and rounded leading edges. First, the PSP method was applied to capture the pressure distributions on the wing suction face; second, the PIV method was used in perpendicular plans to the axis of the model at the positions of chord $x/lo = 0.35–0.9$ for selected angles of attack.

Gursul et al. (2007) studied the concepts of vortex control, which depends on the control of the turbulent flow above the delta wing and enjoys various advantages, such as the lift improvement, drag reduction, and noise attenuation due to the interaction vortex/wing. The control methods include one or more of the following phenomena: the flow separation around the wing, its reattachment on the wing faces, and the vortex bursting. The flow reattachment on the wing faces moves with the increase of the angle of attack and reaches the wing center line at a particular incidence. In these conditions, this particular incidence decreases with the increase of the apex wing. On thicker wings, the vortex bursting appears at very small incidences. The vortex control methods have become increasingly new and diverse. It is useful to consider the physics of the dominant bursting mechanisms which determine the flow control methods.

The state of the art on the various control solutions suggested in the literature was investigated, in order to determine an adequate solution to the aeronautical problems. In order to achieve the objectives previously mentioned, passive (which do not require an external contribution of energy: like the optimization of the wing form) or active (requiring an external contribution of energy: aspiration and blowing of the boundary layer) solutions can be considered. However, most authors did not take into account the interaction of the wing with the wind tunnel walls. The experimental solutions brought by the authors, in particular the blowing and the aspiration of the boundary layer, are the most encouraging solutions of rupture. They made it possible to delay the apex vortex bursting by delaying the layers separation or aspiring it. Nonetheless, they took into account the induced experimental error and the negative effect which these active systems may have on aerodynamics lift and drag. However, no study reported the drag reduction or lift increase of the delta wing through using the concept of privileged angles, on the delta wing geometry, and therefore a great deal of complementary work is still necessary.

Beyond the visualizations and phenomenological analyses existing in the literature, the study suggested here aimed to be a work of quantification, through the parietal pressure distribution. The principal objective was to dissociate the wings with privileged apex from the wings with non-privileged apex and consequently enable us to choose the delta wings.

A series of pressure taps were placed under the principal apex vortex in order to determine the longitudinal distribution of the defect pressure coefficient $-Cp$; finally, the obtained experimental results were confronted. For the three studied delta wings having the following apex angles ($\beta = 75°$, $80°$, and $85°$), the results relating the various aerodynamic coefficients show that the wing with privileged apex angle $\beta = 80°$ has the advantage of presenting the greatest depression values, while the two other delta wings with non-privileged apex angles ($\beta = 75°$ and $85°$) presented lower depression values.

From these results, it can be concluded that it is preferable to use delta wings with privileged apex angle, which allows us better aerodynamic performances. Comparisons between planes provided with delta wings would prove that the wings with privileged apex have advantages related to the stability of the aircrafts and their fuel consumption.

2 Concept of Privileged Angles

Studies in some interesting fields of science, nature, medicine (physiology, anatomy), architecture, physics, and mechanics are more interested in the phenomenological analyses and tend to generalize the existence of privileged angles. In the aerodynamics field, studies based on many visualizations in wind and hydrodynamic tunnels, of flow around bodies of revolutionized and simple delta wings, show that the angle formed by the apex vortex is affected by the wing apex angle value.

The criterion of privileged angle was highlighted, at the atom microscopic scale, and was found by Leray et al. (1972) to exist at a macroscopic scale in the case of Helium II supra fluid flow between the helicoids swirls and their axis. These privileged angles are given by the same following relation (Leray et al. 1985):

$$\cos\beta_{l,m} = \frac{m}{\sqrt{l(l+1)}}, \tag{1}$$

where l and m are integer and with $-l < m < l$.

In Fig. 1, β is the apex angle, α_1 is the angle between the principal vortices, while α_2 is the angle between the secondary vortices.

The above relation (1) enables us to calculate the two main families of privileged angles:

- For $m = l$, we obtain the first family of privileged angles;
- For ($m = 2$, $l \geq 2$), we obtain the second family of privileged angles.

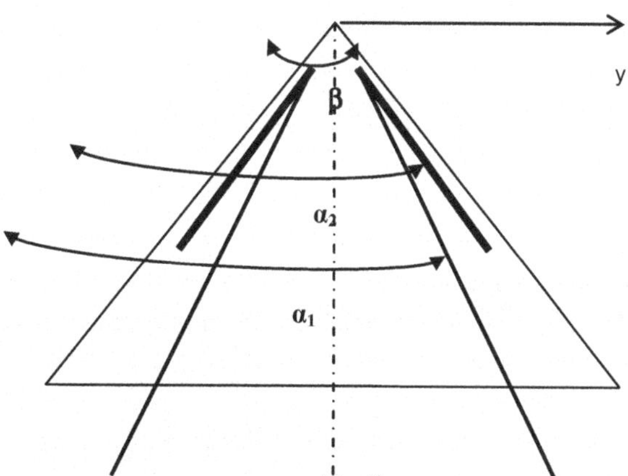

Fig. 1 Vortical system on the upper face of non-privileged delta wing

3 Numerical Method

3.1 Boundary Conditions

A three-dimensional flow simulation around the combinations delta wings–fuselage was carried out using the code Fluent 6.1.22 software (2001). The best way of modeling the test conditions in the wind tunnel was to create a square area around the three-dimensional delta wing with a 300 mm of length and a 300 mm of height. For the rectangular sides of left and right of the field, the flow admission was given by "Velocity Inlet" and the type of "Exit" was adopted. The four other sides were considered as walls conditions in the section of the wind tunnel. Calculations were carried out for the flow nominal velocity $V_0 = 20$ m/s.

3.2 Mathematical Formulation

The Spalart–Allmaras turbulence model with one equation was used during simulations. The solution variables of the instantaneous Navier–Stokes equations were decomposed into the mean and fluctuating components (for the velocity: $u_i = \bar{u}_i + u_i'$). Substituting their expressions in the instantaneous continuity and momentum equations, and taking a time average, yields they can be written in Cartesian tensor form as

$$\frac{\partial \rho}{\partial t} + \frac{\partial(\rho u_i)}{\partial x_i} = 0 \tag{2}$$

$$\frac{\partial(\rho u_i)}{\partial t} + \frac{\partial(\rho u_i u_j)}{\partial x_j} = -\frac{\partial p}{\partial x_j} + \frac{\partial}{\partial x_j}\left[\mu\left(\frac{\partial u_i}{\partial x_j} + \frac{\partial u_j}{\partial x_i} - \frac{2}{3}\delta_{ij}\frac{\partial u_i}{\partial x_i}\right)\right] + \frac{\partial}{\partial x_j}\left(-\rho\overline{u_i'u_j'}\right)$$
$$(i = 1, 2, 3). \tag{3}$$

It is worth noting, now, that some additional terms that represent the effects of turbulence appear. These Reynolds stresses, $-\rho\,\overline{u_i'u_j'}$, must be modeled in order to close Eq. (3). A common method employs the Boussinesq hypothesis to relate the Reynolds stresses to the mean velocity gradients:

$$-\overline{u_i'u_j'} = \frac{\mu_t}{\rho}\left(\frac{\partial u_i}{\partial x_j} + \frac{\partial u_j}{\partial x_i}\right) - \frac{2}{3}\left(k + \frac{\mu_t}{\rho}\frac{\partial u_i}{\partial x_i}\right)\delta_{ij} \tag{4}$$

The Boussinesq hypothesis is used in the Spalart–Allmaras model. The advantage of this approach is the relatively low computational cost associated with the computation of the turbulent viscosity μ_t. In the case of Spalart–Allmaras model, only one additional transport equation (representing turbulent viscosity) is solved. Note that since the turbulence kinetic energy k is not calculated in the Spalart–Allmaras model, the last term in Eq. (4) is ignored when estimating the Reynolds stresses.

3.3 Grid

The type of grid element employed, schematized on Fig. 2, is triangular. Besides, an unstructured grid was applied. The majority of the significant properties of the flow to be reproduced are in the proximity of the wing surface. Consequently, the

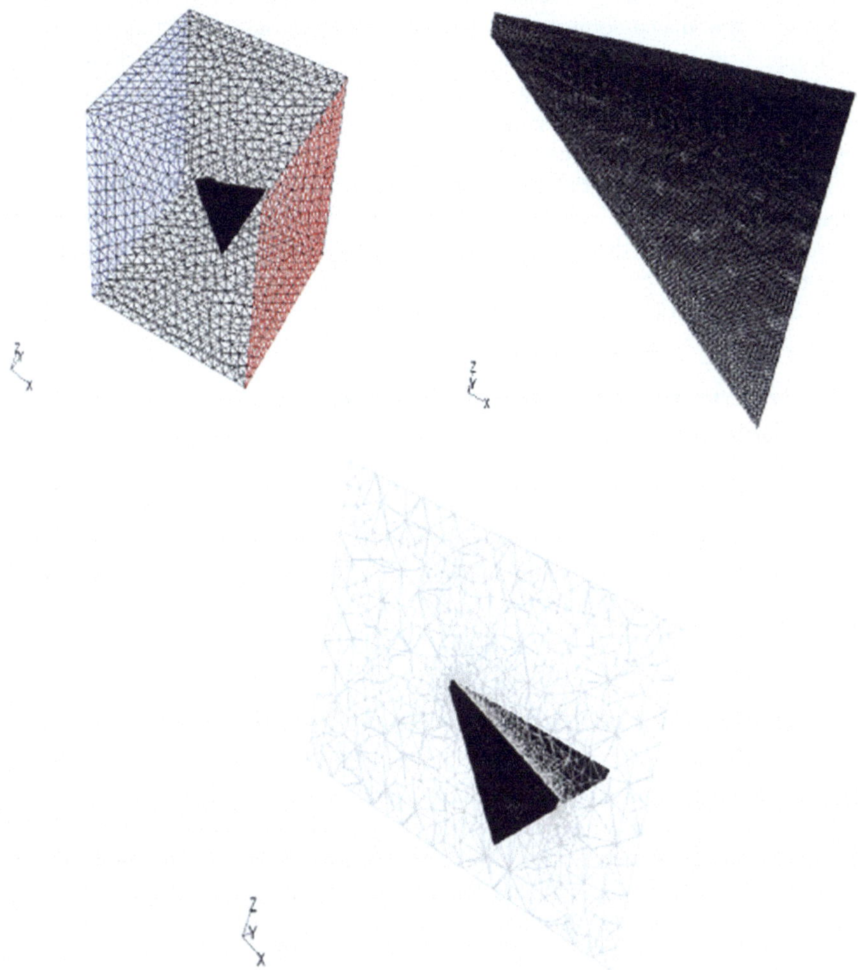

Fig. 2 Meshed geometry example (wing and delta wing–fuselage combination)

meshed field of calculation was refined close to the suction face and the wing faces. However, it was maintained gross in the rest areas of the field to decrease the calculation time.

4 Numerical Results

4.1 Defect Pressure Coefficient −Cp Contours

Figures 3, 4, and 5 present the distribution of the defect pressure coefficient −Cp contours obtained with the numerical simulation, at the suction face of the various studied delta wings. The vortex structure developing at the suction face of the delta wings was reproduced compared to the visualizations results of Benkir (1990).

In Fig. 4, we remark that the angle between the apex vortices direction is very affected by the presence of the fuselage diameter $d = 20$ mm. This fact is valuable for the delta wing with apex angles $\beta = 75°$, 80°, and 85°. The value of −Cp is important for the case of wing without fuselage for all the studied apex.

In Fig. 5, we see that the apex vortices move toward the delta wing leading edges and an interaction zone between the delta wing and the fuselage appeared at the junction point for the two considered fuselages of diameters $d = 20$ mm and $d = 30$ mm.

4.2 Transverse Evolution of Defect Pressure Coefficient −Cp

Figures 6 and 7 show that according to the four cross sections considered at the top of the delta wings, we note that the apex vortex position is detected by the maximum value of −Cp. The maximum value of −Cp decreases when a cylindrical fuselage of diameter d is introduced.

4.3 Role of the Privileged Apex Angle $\beta = 80°$

The longitudinal evolution of the defect pressure coefficient −Cp which is shown in Fig. 8 proves clearly that there is an effect of the privileged angle apex $\beta = 80°$ and the value of the most significant −Cp is reached around $r/lo = 0.3$ for the Reynolds number $1.23 \times 10^{+5}$. Through the longitudinal evolution of the defect pressure coefficient −Cp, under the principal apex vortex, we can easily notice the maximum

Fig. 3 Contours of −Cp at the suction face of delta wings without fuselage at V_0 = 20.3 m/s and an incidence angle i = 15°

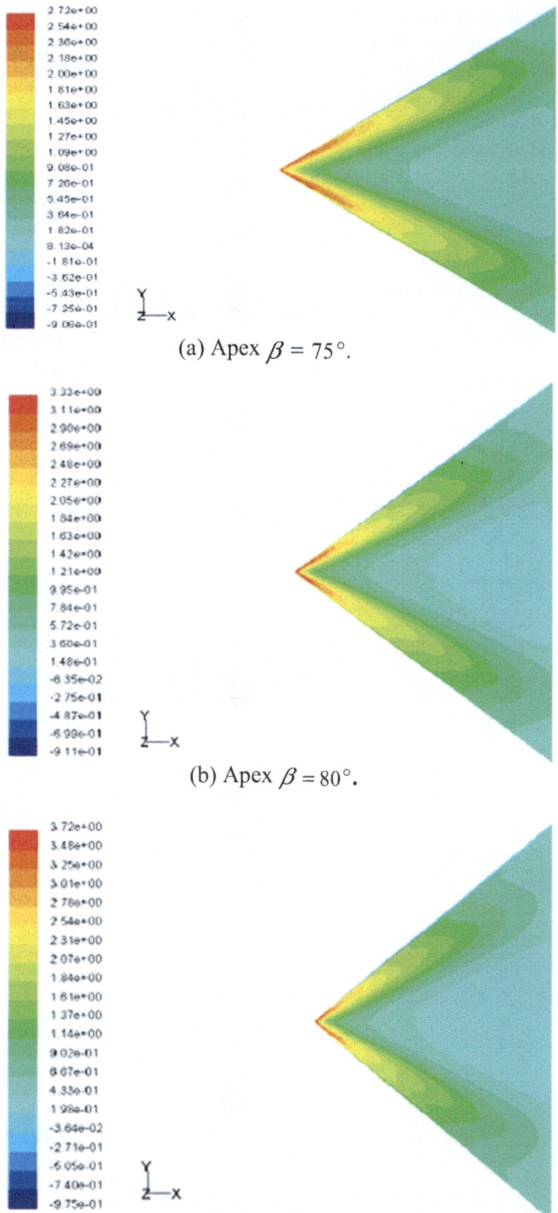

(a) Apex $\beta = 75°$.

(b) Apex $\beta = 80°$.

(c) Apex $\beta = 85°$.

Fig. 4 Contours of −Cp at the suction face of the combinations delta wings–fuselage with a diameter $d = 20$ mm at $V_0 = 20.3$ m/s and an incidence angle $i = 15°$

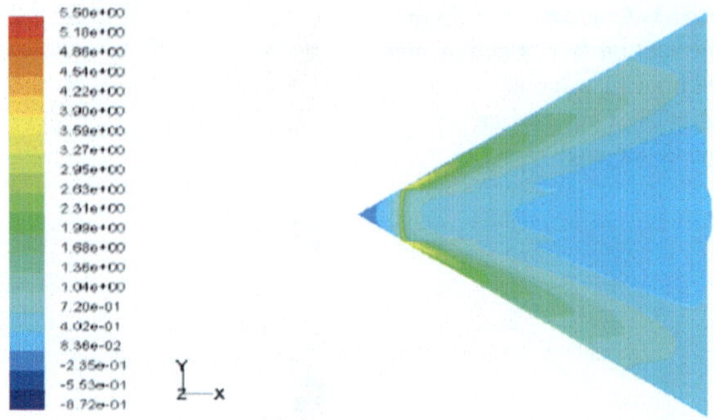

(a) Apex $\beta = 75°$.

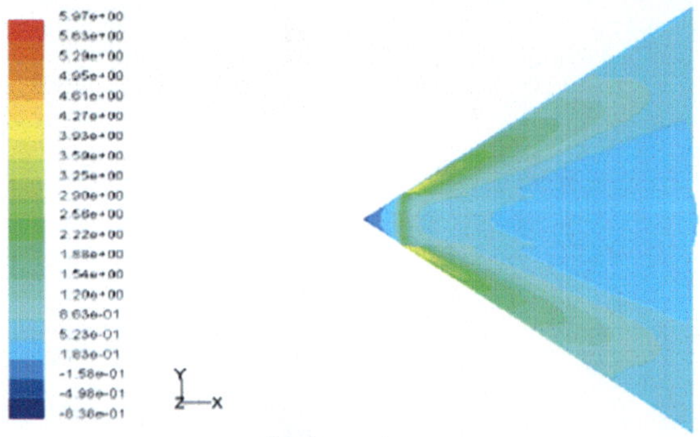

(b) Apex $\beta = 80°$.

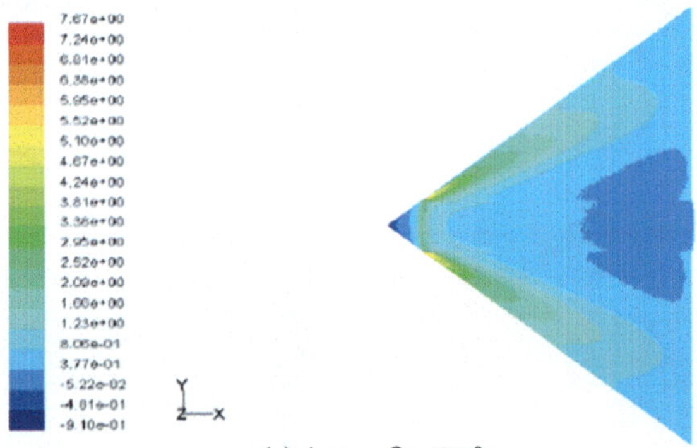

(c) Apex $\beta = 85°$.

Fig. 5 Contours of −Cp at the suction face of the combinations delta wings–fuselage with a diameter d = 30 mm at V_0 = 20.3 m/s and an incidence angle i = 15°

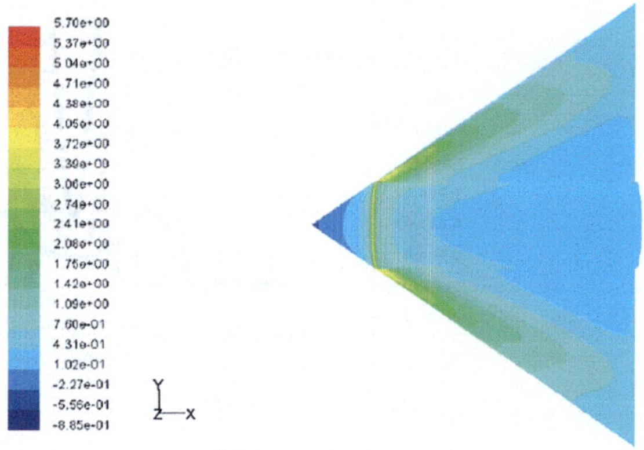

(a) Apex $\beta = 75°$.

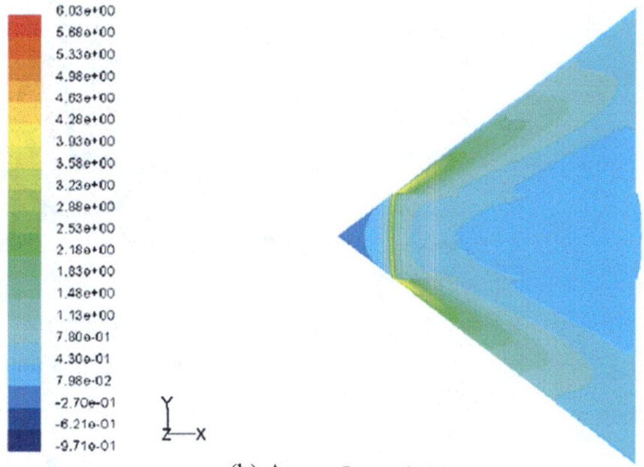

(b) Apex $\beta = 80°$.

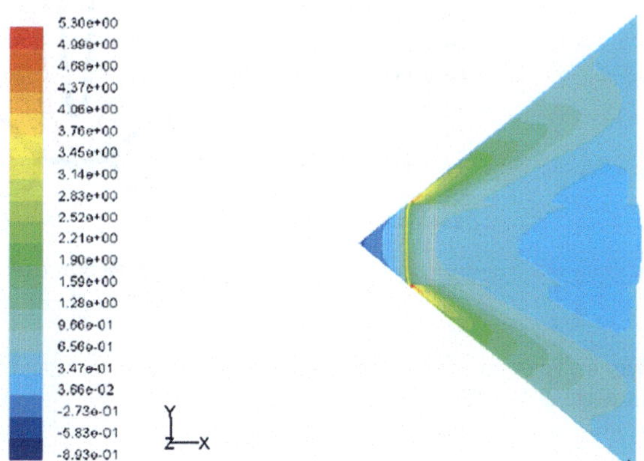

(c) Apex $\beta = 85°$.

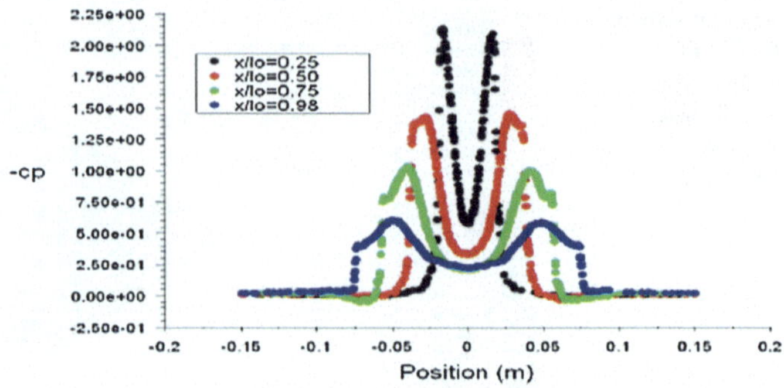

(a) Wing with apex angle β=75°.

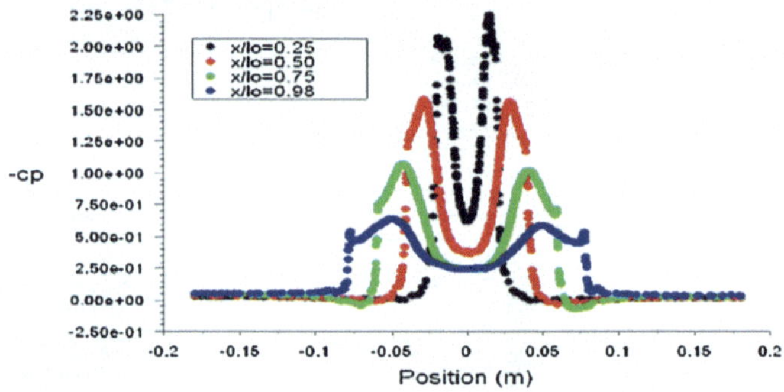

(b) Wing with apex angle β=80°.

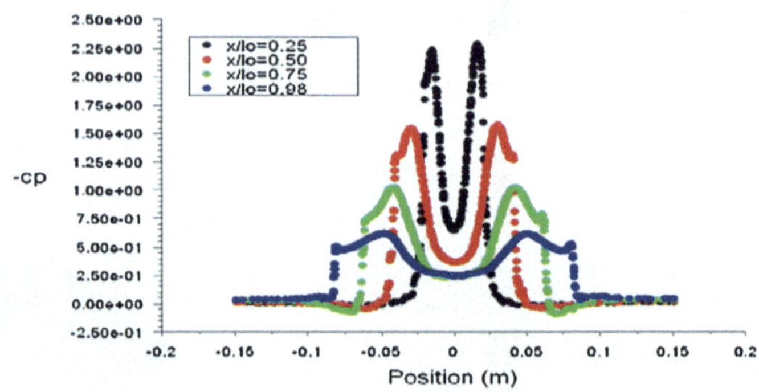

(c) Wing with angle of apex β=85°.

Fig. 6 Transverse evolution of numerical −Cp at the suction face of the delta wings without fuselage at V_0 = 20.3 m/s and incidence i = 15°

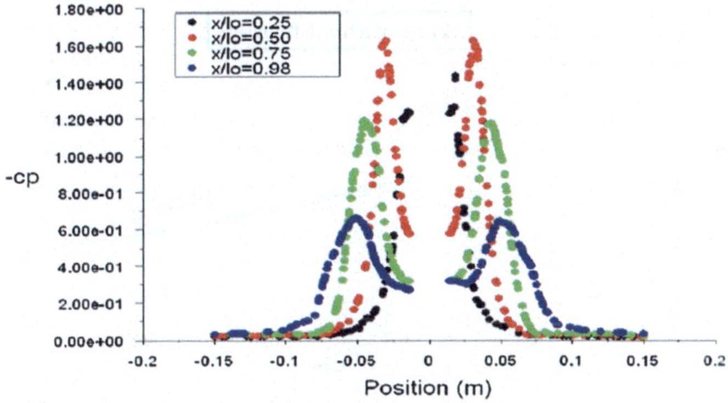

(a) Wing with apex angle β=75°.

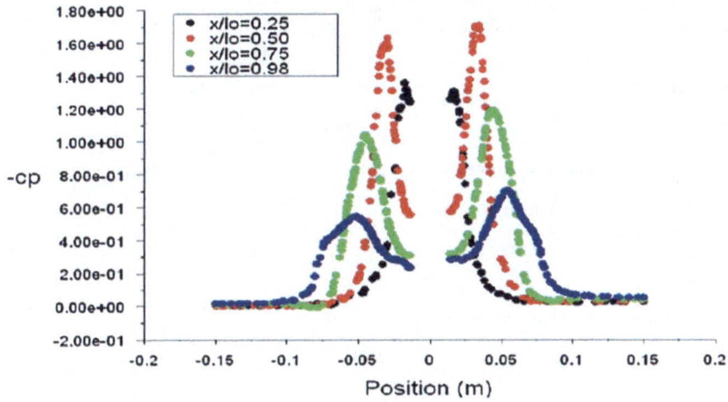

(b) Wing with apex angle β=80°.

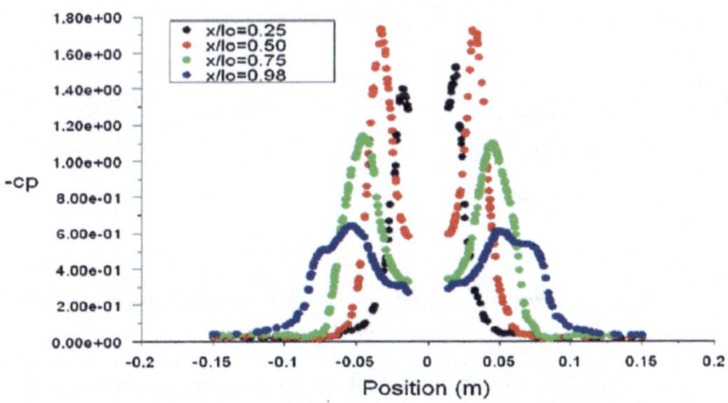

(c) Wing with angle of apex β=85°.

Fig. 7 Transverse evolution of numerical −Cp at the suction face of the combinations delta wings–fuselage d = 30 mm at V_0 = 20.3 m/s and an incidence angle i = 15°

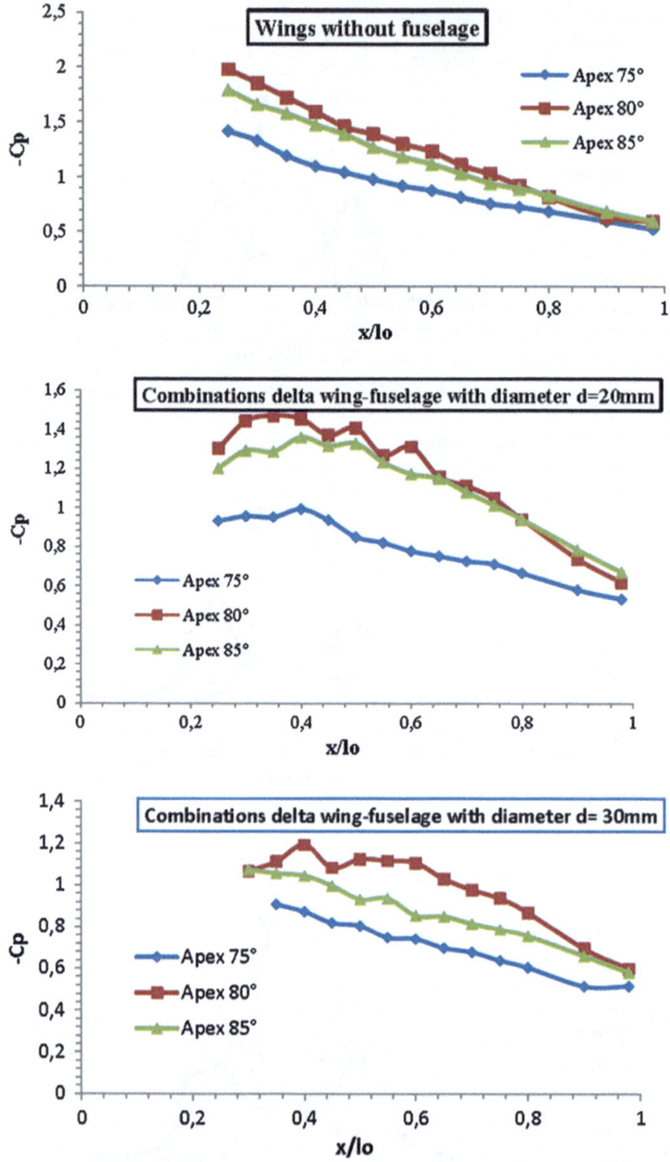

Fig. 8 Longitudinal evolution of the numerical defect pressure coefficient −Cp under the principal apex vortex of delta wings without fuselage and combinations delta wing–fuselage at $V_0 = 20.3$ m/s and an incidence angle $i = 15°$

numerical values of −Cp for the wing with privileged angle apex $\beta = 80°$. This fact is valuable for the wings without fuselage or for the combinations delta wing–fuselage (Fig. 8).

4.4 Fuselage Diameter Effects

Figure 9 shows the effect of the fuselage diameter on the longitudinal evolution of the numerical −Cp under the apex vortex at $V_0 = 20.3$ m/s and an incidence angle $i = 15°$. According to these results, it can be noted that for $x/lo < 0.6$ corresponding to the vicinity immediate of the delta wing apex, the maximum values of −Cp are obtained with the wing without fuselage. Then, the values corresponding to the combination delta wings–fuselage appear with diameter $d = 20$ mm and lastly those corresponding to the combination with $d = 30$ mm. Beyond $x/lo = 0.60$, i.e., when we move toward the wing trailing edge, the same order is preserved with a tendency of confusion of the considered curves (Fig. 9).

According to these results, we can deduce that we have the same evolution of the numerical results concerning the role of privileged apex angle $\beta = 80°$ (Fig. 8) for the wings without fuselage and for the combinations delta wing–fuselage. The effect of the diameter presence is the same concerning the three studied apex cases (Fig. 9).

5 Comparison with Experimental Results

The current way used to ensure the quality of the numerical simulations is the calculation of the aerodynamic characteristics such as pressure distribution and forces, and compares them with the experimental results obtained in the wind tunnel.

The numerical and the experimental results are confronted in the same graphs (Figs. 10 and 11) for better elucidating the difference. According to these figures, it was noted that for the wings without fuselage, the defect pressure coefficient read on the experimental and numerical curves are in good agreement, and the difference between these two last curves is due to the resolution of the Navier–Stokes equations with simplifying assumptions. For the case of the combination delta wing–fuselage, the difference between the experimental and the numerical curves increases as we approach the wing apex.

For the combinations wing–fuselage (Fig. 11), the two curves converge when we move away from the apex of the wing. In fact, the difference between the two curves is more significant at the vicinity of the apex and becomes negligible when we approach the trailing edge of the wing.

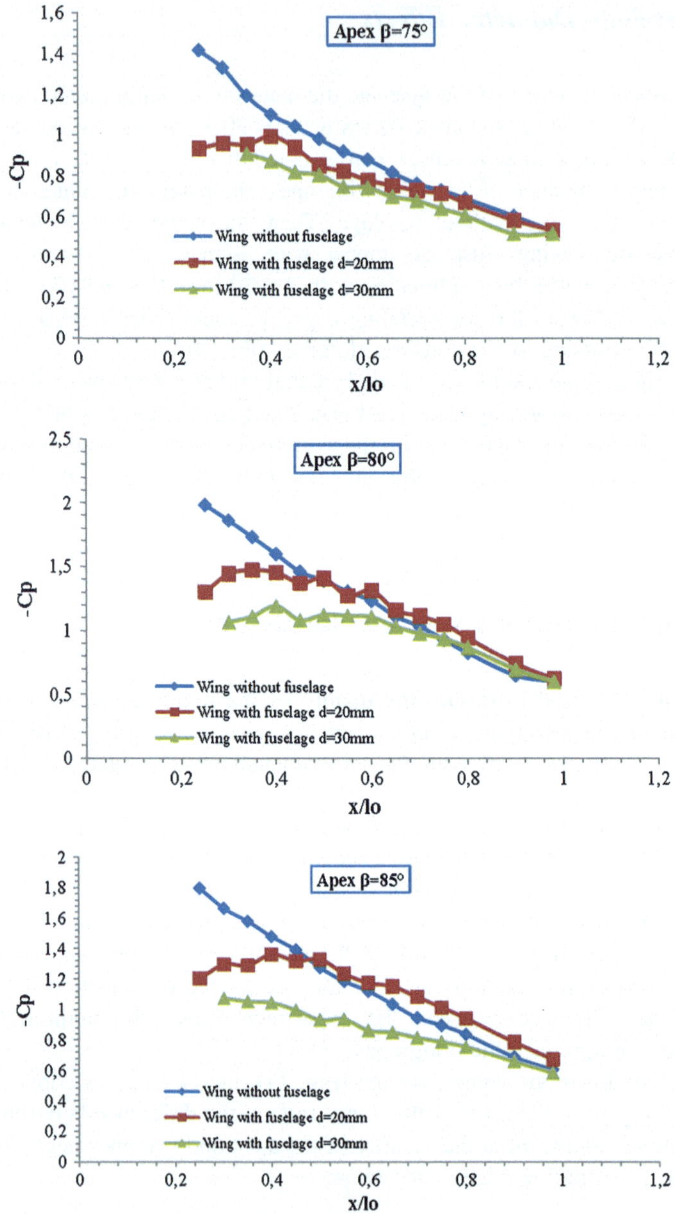

Fig. 9 Fuselage diameter effect on the longitudinal evolution of the numerical −Cp under the apex vortex at V_0 = 20.3 m/s and an incidence angle i = 15°

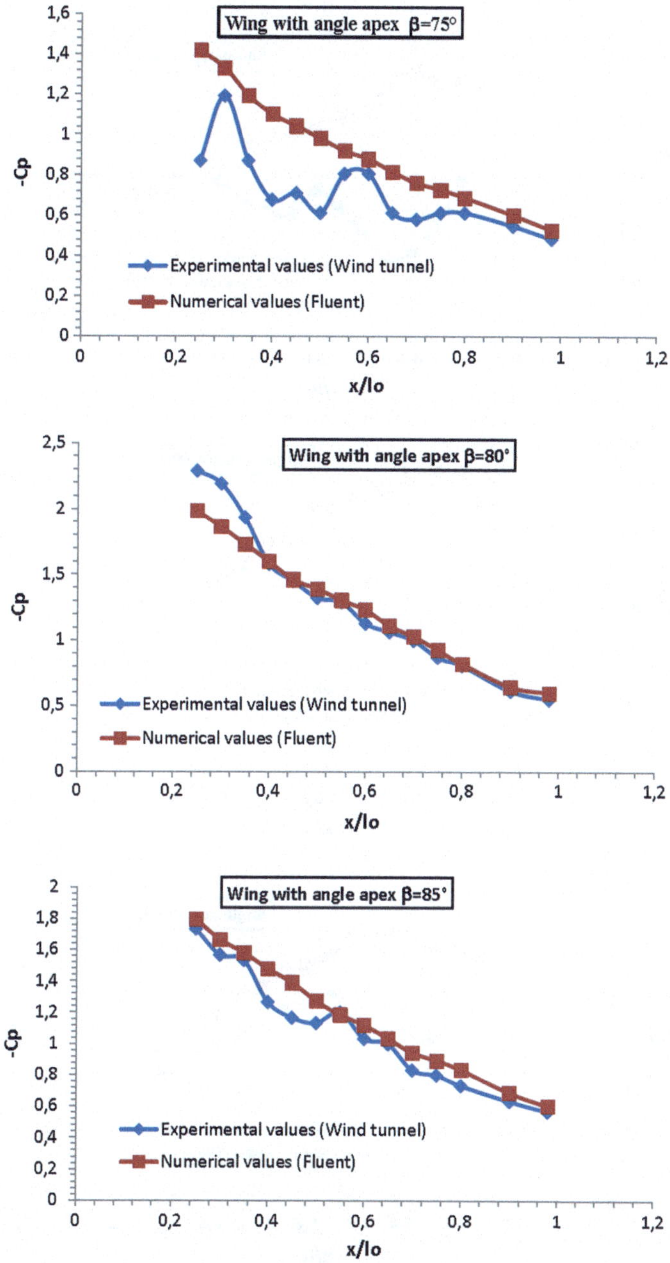

Fig. 10 Comparison between the numerical and experimental values of −Cp, under the principal apex vortex, for the delta wings without fuselage at $V_0 = 20.3$ m/s and an incidence angle $i = 15°$

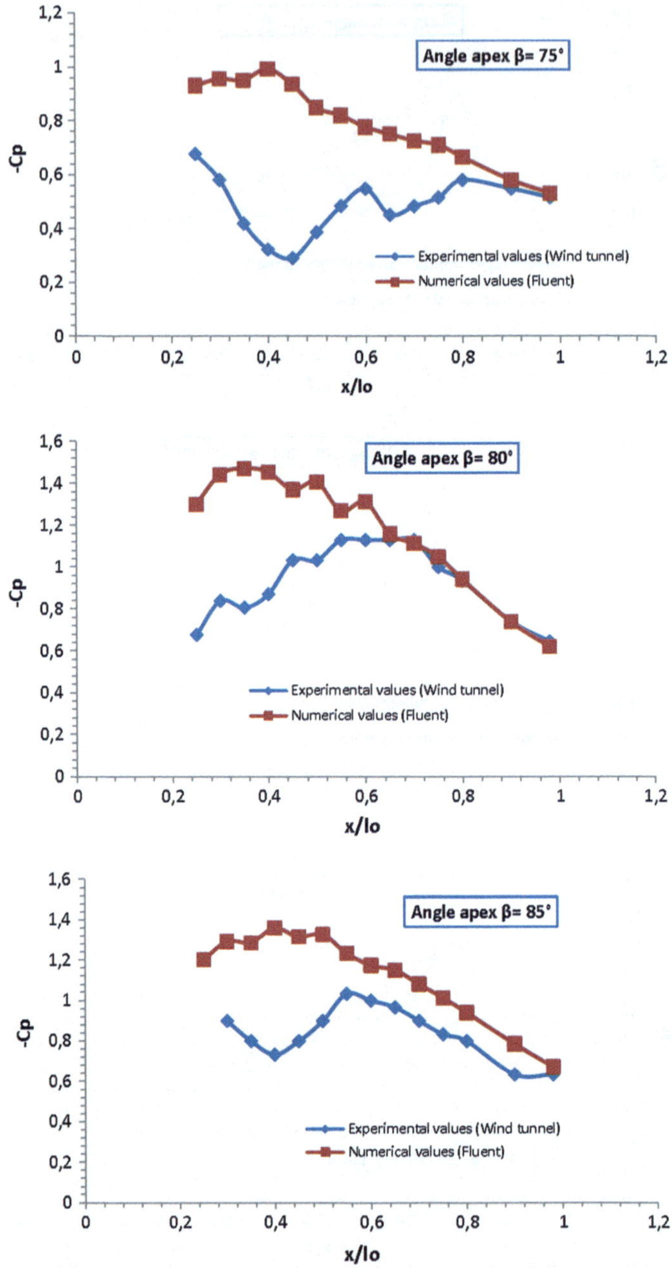

Fig. 11 Comparison between the numerical and experimental values of −Cp, under the principal apex vortex, for the various combinations delta wing–fuselage with diameter $d = 20$ mm at $V_0 = 20.3$ m/s and an incidence angle $i = 15°$

6 Conclusions

The discussions of the various experimental and numerical results were achieved by taking into account the geometrical and kinematics parameters, and we can, thus, conclude that

(1) the increase of the incidence always generates an increase in the coefficient of depression −Cp, for all the studied wings as long as the incidence of the wing is lower than the critical incidence (incidence of vortex bursting).
(2) at the high attack angles ($i > 25°$), the flow at the extrados of the delta wings without fuselage was mainly asymmetrical, and the part close to the apex shows high values of −Cp. The effects of the Reynolds number on the flow around the delta wings without fuselage are not significant. The vortex breakdown incidence of the apex vortex is reached at lower incidences for the wings with apex angles $\beta = 80°$ and 85° ($i_{vb} = 22°–25°$), compared to the wing with angle apex $\beta = 75°$ ($i_{vb} = 30°$). This fact is valid for the all used velocities flow.
(3) the depression values corresponding to the delta wings with privileged apex angles are more significant than those obtained for the wings with non-privileged apex angles $\beta = 75°$ and 85°, which enables us to reach one of the objectives fixed at the beginning of our work consisting on studying the role of the privileged angle on flow around delta wings.

A systematic experimental study was carried out for three configurations of delta wing–fuselage at various attack angles and flow velocities. We observe, on the combinations delta wing–fuselage, the effects of the privileged angle $\beta = 80°$ until the attack angle $i = 22°–25°$, where the apex vortex breakdown takes place, downstream from $r/lo = 0.6$; a slight change of flow was recorded through the aerodynamic characteristics evolution.

At high attack angles ($i > 25°$), the flow around the combinations delta wing–fuselage was mainly asymmetrical, and the fuselage induces the asymmetry of vortex or its bursting or both. The effects of the Reynolds number on the flow around the combination delta wing–fuselage are insignificant.

Comparing the numerical and experimental results, we have noticed a better reproduction of the apex vortices at the suction face of the studied delta wings and a very good agreement between these results for the studied wings and combination delta wings–fuselage.

Simulation of airflow around the delta wing and around the various studied combinations enabled us to well understand the physical phenomenon through the obtained contours of defect pressure coefficient and velocity.

In addition, by analyzing the flow around the various combinations, it has been noted that the role of the privileged angle apex $\beta = 80°$ appears for the delta wings without fuselage and with fuselage of diameters $d = 20$ mm and $d = 30$ mm.

So far, the obtained numerical results have evolved in the good direction: no contradictions were remarked between the achieved simulations and the measurements really undertaken in the wind tunnel.

Nomenclature

Cp	Pressure coefficient
g	Acceleration of gravity
i	Angle of attack (AOA°): incidence (°)
i_{cr}	Critical incidence (°)
i_{vb}	Vortex breakdown incidence (°)
k	Kinetic energy
l_o	Wing chord (m)
L	Wing span (m)
O^{xyz}	Cartesian coordinates axis
O^{x}	Median axis on the wing surface
O^{y}	Transverse axis on the wing surface
O^{z}	Vertical axis
r	Polar coordinate of the pressure taps (m),
S	Wing surface (m^2),
u_i	Velocity component (m/s), (i = 1, 2, 3),
$\bar{u}_i$	Mean velocity component,
u'_i	Fluctuating velocity component,
V_0	Wind tunnel velocity (m/s).

Greek Letters

ρ	Air density (kg/m^3),
ρ_H	Oil density (kg/m^3),
α_1	Angle between the principal vortices directions (°),
α_2	Angle between the secondary vortices directions (°),
β	Apex angle (°),
δ_{ij}	Index of Kronecker,
μ_t	Turbulent viscosity (Ns/m^2),
$\lambda = \frac{L^2}{S} = 4tg\left(\frac{\beta}{2}\right)$	Aspect ratio,
θ	Polar coordinate of the pressure taps (°),
φ	Sweep angle (°).

References

Benkir M (1990) Persistance et destruction des structures tourbillonnaires concentrées en particulier au dessus d'ailes delta: Critères angulaires de stabilité aux écoulements. Université de Valenciennes, Thèse de Doctorat

Brodetsky MD, Krause E, Nikiforov SB, Pavlov AA, Kharitonov AM, Shevchenko AM (2001) Evolution of vortex structures on the leeward side of a delta wing. J App Mech Tech Phy 42(2): 243–254

Brown CE, Michael WH (1955) On slender delta wings with leading-edge separation. NACA Technical note 3430

Cummings RM, Rizk YM, Schiff LB, Chaderjian NM (1990) Navier-Stokes predictions of the flow field around the F-18 (HARV) wing and fuselage at large incidence. AIAA 90-0099, presented at 28th aerospace sciences meeting, Reno, NV, USA

Ekaterinaris JA, Schiff LB (1990) Vortical flows over delta wings and numerical prediction of vortex breakdown. AIAA-90-0102, presented at 28th aerospace sciences meeting, Reno, NV, USA

Gursul I, Vardaki E, Margaris P, Wang Z (2007) Control of wing vortices. In: King R (ed) Active flow control, NNFM 95. Springer, Berlin, pp 137–151

Hawk JD, Barnet RM, O'neil PJ (1990) Investigation of high angle of attack vortical flow over delta wings, AIAA-90-0101. Presented at 28th aerospace sciences meeting, Reno, NV, USA

Konrath R, Klein C, Schroder A, Kompenhans J (2008) Combined application of pressure sensitive paint and particle image velocimetry to the flow above a delta wing. Exp Fluids 44:357–366 (Research Article)

Laidlaw WR, Halfman RL (1956) Experimental pressure distributions on oscillating low aspect ratio wings. J Aeronaut Sci 23:117–124

Lambourne NC, Bryer DW (1962) The bursting of leading-edge vortices-some observations and discussion of the phenomenon. Aeronautical Research Council, R&M, No. 3282

Lambourne NC, Bryer DW, Maybrey JFM (1964) The behavior of leading edge vortices over a delta wing following a sudden change of incidence. Reports & Memoranda No. 3645, Aeronautical Research Council, London, UK

Lang N (2004) Investigation of the supersonic flow field around a delta wing using particle-image-velocimetry. Aerodynamisches Institut, RWTH-Aachen, Wüllnerstr. zw. 5 u. 7, D-52062 Aachen, Germany

Leray M, Deroyon MJ, Francois M, Vidal F (1972) New results on the vortex lattice. Proceedings of the 13th international conference on low temperature, Physics LT 13 Boulder, Colorado, USA

Leray M, Deroyon JP, Deroyon MJ, Minair C (1985) Critères angulaires de stabilité d'un tourbillon hélicoïdal ou d'un couple de tourbillons rectilignes, rôle des angles privilégiés dans l'optimisation des ailes, voiles, coques des avions et navires. Association technique maritime et aéronautique, France

Lowson MV (1963) The separated flow on slender wings in unsteady motion. Reports & Memoranda No. 3448, Aeronautical Research Council, London, UK

Parker AG (1976) Aerodynamic characteristics of slender wings with sharp leading edges—a review. J Aircr 13(3):161–168

Randall DG (1966) Oscillating slender wings with leading-edge separation. Aeronaut Q XVII:311–331

User Manual Fluent 6.1.22 (2001) Fluent Inc.

Wang JJ, Li QS, Liu JY (2003) Effects of a vectored trailing edge jet on delta wing vortex breakdown. Exp Fluids 34:651–654

Woodgate L, Halliday AS (1971) Measurements of the oscillatory pitching-moment derivatives on a series of three delta wings in incompressible flow. Reports & Memoranda No. 3628, parts 1–4, Aeronautical Research Council, London, UK

Numerical Simulation of the Overlap Effect on the Turbulent Flow Around a Savonius Wind Rotor

Sobhi Frikha, Zied Driss, Hedi Kchaou and Mohamed Salah Abid

1 Introduction

Renewable energies are clean and inexhaustible and therefore environmentally friendly. They are provided by the sun, the wind, the waves, or the heat of the earth's core. A variety of methods are used to convert these renewable resources into electricity. Wind power is one of the fastest growing renewable energy technologies, and wind turbines have been developed to generate electricity from the kinetic power of the wind. Wind turbines can rotate around either a horizontal or a vertical axis. Savonius wind turbines are a type of vertical-axis wind turbine that operates due to drag forces on their buckets. The Savonius wind rotor has the advantage of being compact, economic, and aesthetic. In addition, it has good starting characteristics, operates at relatively low operating speeds, and can receive wind from any direction. For several years, many studies have significantly improved the performance of Savonius rotors. Aldos (1984), for instance, studied the Savonius rotor power increase by allowing the rotor blades to swing back when on the upwind side. He reported a power augmentation of the order of 11.25% with the increase in C_p from 0.015 to 0.17. Ushiyama and Nagai (1988) tested several

S. Frikha (✉) · Z. Driss · H. Kchaou · M. S. Abid
Laboratory of Electro-Mechanic Systems (LASEM), National School of Engineers of Sfax (ENIS), University of Sfax, Road Soukra Km 3.5, B.P. 1173, 3038 Sfax, Tunisia
e-mail: Frikha_sobhi@yahoo.fr

Z. Driss
e-mail: Zied.Driss@enis.rnu.tn

H. Kchaou
e-mail: hedi.kchaou@ipeis.rnu.tn

M. S. Abid
e-mail: MohamedSalah.Abid@enis.rnu.tn

Z. Driss et al. (eds.), *CFD Techniques and Thermo-Mechanics Applications*,
https://doi.org/10.1007/978-3-319-70945-1_8

parameters of the Savonius rotor including the gap ratio, the aspect ratio, the number of cylindrical buckets, the number of stages, the end plate effects, the overlap ratio, and the bucket design. The highest efficiency of all the tested configurations was 24% for a two-stage, two-bucket rotor. Grinspan et al. (2001) developed a new blade shape with a twist for the Savonius rotor. They obtained a maximum power coefficient of 0.5 with this model. Shikha et al. (2003) used a convergent nozzle at the front of the Savonius rotor. The nozzle increases the power at lower wind speeds and prevents the negative torque. Menet and Bourabaa (2004) tested different configurations of the Savonius rotor and found that the best value of the static torque coefficient is obtained for an incidence angle equal to $\theta = 45°$ and a relative overlap equal to $e/d = 0.24$. They compared their numerical results with those obtained by Blackwell et al. (1978), and a good agreement was obtained. Saha and Rajkumar (2005) compared the performance of a bladed metallic Savonius rotor to a conventional semicircular blade with no twists. The twist produced good starting torque and larger rotational speeds, and gives an efficiency of 0.14. The best torque was obtained with twisted blades by an angle $\alpha = 12.5°$. Kamoji et al. (2009) investigated the performance of modified forms of conventional rotors with and without central shaft between the end plates. Khan et al. (2009) tested different blade profiles of a Savonius rotor both in tunnel and natural wind conditions, and they varied the overlap. The highest C_p of 0.375 was obtained for a blade profile of S-section Savonius rotor at an optimum overlap ratio of 30%. Driss and Abid (2012) conducted a computational fluid dynamic study to present the local characteristics of the turbulent flow around a Savonius wind rotor. They compared their numerical results with experimental results, and a good agreement was obtained. Driss et al. (2015) compared different rotor designs characterized by the bucket angles equal to $\psi = 60°$, $\psi = 75°$, $\psi = 90°$, and $\psi = 130°$. It has been noted that the depression zones increased with the bucket arc angle increase. The acceleration zone, where the maximum velocity values are recorded, is formed on the convex surface of the rotor bucket and gets greater on the bucket arc. The wakes characteristics of the maximum turbulent values are more developed with the increase in the bucket arc angle. Akwa et al. (2012) studied the influence of the buckets overlap ratio of a Savonius wind rotor on the averaged moment and power coefficients by changing the rotor geometry. They noticed that the maximum device performance occurs for buckets overlap ratios with values close to 0.15. Rogowski and Maroński (2015) studied the aerodynamic efficiency of the Savonius rotor using computational methods of fluid dynamics. The study has demonstrated that the CFD methods confirm the experimental results and can be used to optimize the Savonius rotor buckets shape. Sharma and Gupta (2013) have studied the performance of a three-bucket Savonius rotor by the CFD software Fluent 6. The flow behavior around the rotor was also analyzed with the help of pressure, velocity, and vorticity contours, for different overlap ratios. Choudhury and Saraf (2014) have analyzed the flow characteristics of two-bladed Savonius rotor with 2D and 3D analyses using CFD ANSYS Fluent software. They also studied the static pressure, the velocity, the vorticity, and the turbulent kinetic energy. According to their results, the drag and the torque coefficients are maximum, respectively, at 0° and 30° rotor blade angles, and the vorticity and the turbulent kinetic energy show the maximum value at 30° rotor blade angle. The above results on Savonius rotors give different optimum geometries. However, a paucity has been noted in the study of the aerodynamic characteristics of the Savonius rotor. In this chapter, we are interested in studying the flow around a Savonius wind rotor. Thus, numerical simulations of the turbulent flow have been developed using a CFD code, and the effect of the overlap on the aerodynamic characteristics of the flow has been investigated.

2 Geometric Parameters and Boundary Conditions

The examined Savonius rotor consists of two half-cylinder buckets of diameter $d = 0.1$ m (Fig. 1). The shaft diameter is equal to $e' = 10$ mm. In this work, we have studied different overlap values equal to $(e\text{-}e')/d = 0$, $(e\text{-}e')/d = 0.1$, $(e\text{-}e')/d = 0.24$, and $(e\text{-}e')/d = 0.3$. For the boundary conditions, we took a value of $V = 9.95\ \text{m s}^{-1}$ for the inlet velocity. For the pressure outlet, a value of $p = 101{,}325$ Pa was considered (Fig. 2).

3 Numerical Results

3.1 Velocity Field

Figure 3 presents the distribution of the velocity field in the computational domain with a zoom around the Savonius wind rotor for different overlap values. A similarity in the distribution of the velocity field seems to be clear enough from these results. In fact, at the entry of the computational domain, the flow appears uniform and the velocity is equal to $V = 9.95\ \text{m s}^{-1}$. Upstream of the Savonius rotor, there is a flow deceleration at both concave and convex surfaces of the rotor. However, at the external attack zone, there is an increase in the velocity values.

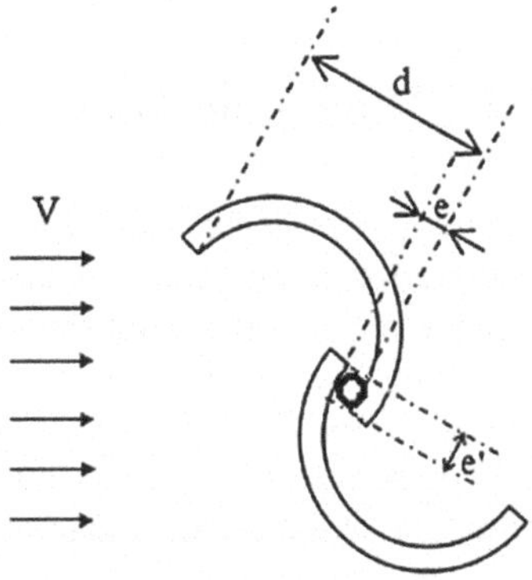

Fig. 1 Savonius rotor

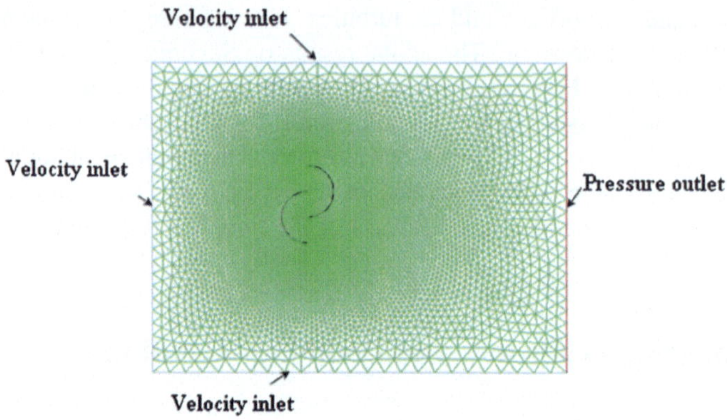

Fig. 2 Boundary conditions

This increase generates an acceleration zone of the velocity. Behind the rotor, there is a rapid deceleration in the velocity field and the formation of two recirculation zones downstream of the two buckets. These zones extend to the limits of the computational domain. By comparing these results together, the internal overlap is observed to have a direct effect on the velocity fields distribution. Indeed, in the case of a nonzero overlap ($(e\text{-}e')/d \neq 0$), the flow strikes the concave surface of the upper rotor and seeps through the gap with a quite high velocity in the lower attack zone of the upper bucket of the Savonius rotor. Then, the flow continues along the concave surface of the lower bucket of the rotor. The reduction of the overlap causes an increase in the velocity value. When the overlap is null, which corresponds to an overlap equal to $(e\text{-}e')/d = 0$, the flow is stopped at the rotor axis. The fluids particles change their direction and seep through the upper attack zone of the upper bucket. Globally, the maximum value of the velocity was noticed to decrease while increasing the overlap. For the different overlap values equal to $(e\text{-}e')/d = 0$, $(e\text{-}e')/d = 0.1$, $(e\text{-}e')/d = 0.24$, and $(e\text{-}e')/d = 0.3$, the maximum values of the velocity are equal, respectively, to $V = 16.7$ m s^{-1}(Fig. 3a), to $V = 15.5$ m s^{-1} (Fig. 3b), to $V = 14.9$ m s^{-1} (Fig. 3c), and to V = 14.7 m s^{-1} (Fig. 3d).

3.2 *Mean Velocities*

Figure 4 presents the distribution of the mean velocity in the computational domain with a zoom around the Savonius wind rotor for different overlap values. According to these results, the mean velocity is noted to be equal to 9.95 m s^{-1} at the entry of the computational domain. At the two buckets, the mean velocity decreases and reaches zero. Furthermore, an occurrence of breakpoints was observed on the convex surface of the lower bucket and the concave surface of the upper bucket. However, at the external attack zone of the two buckets, there is a rapid increase in

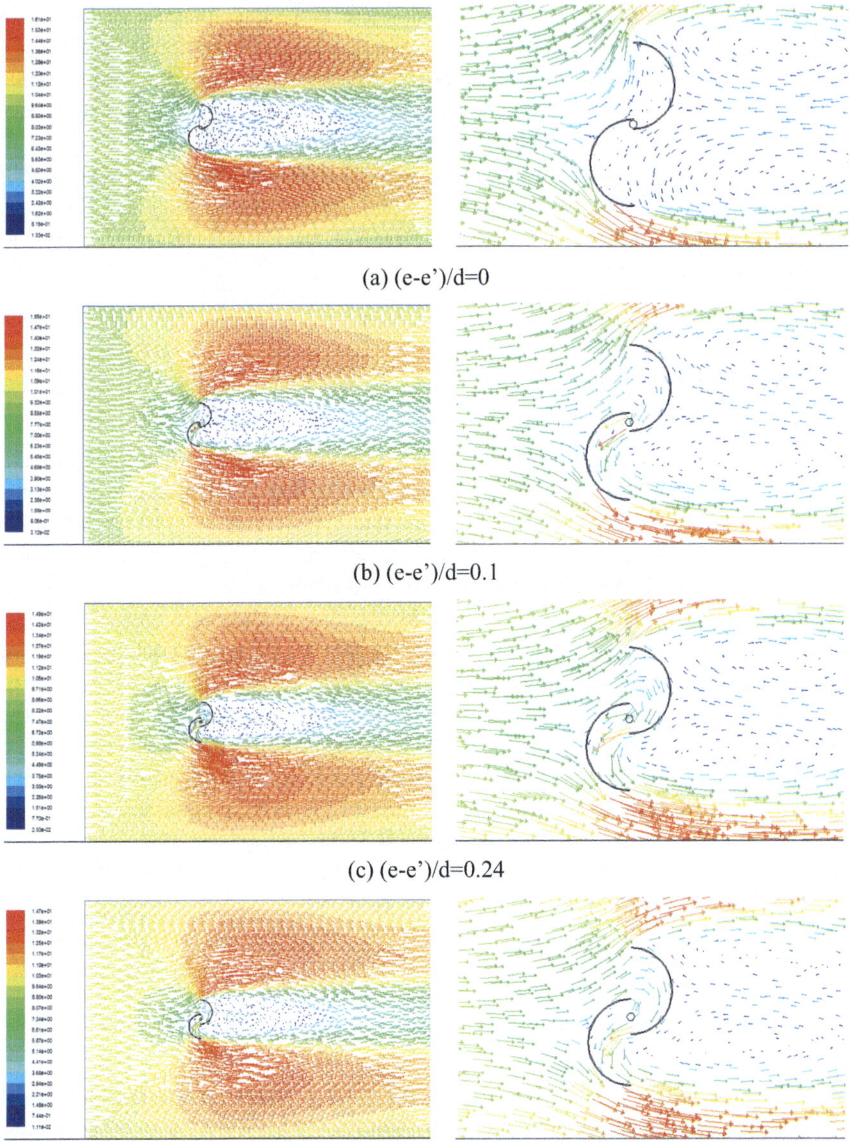

(a) (e-e')/d=0

(b) (e-e')/d=0.1

(c) (e-e')/d=0.24

(d) (e-e')/d=0.3

Fig. 3 Distribution of the velocity field

the mean velocity values. From both sides, there is a wake characteristic of the maximum values of the mean velocity, which extends to the limits of the computational domain. Behind the rotor, there is a rapid decrease in the mean velocity values. By comparing these results together, the internal overlap is noted to have a direct effect on the mean velocity. Indeed, there is a rapid increase in the mean velocity in the internal attack zone in the case of a nonzero overlap ($(e\text{-}e')/d \neq 0$). This fact is caused by the infiltration of the air stream between the two buckets. Indeed, it has been noted that the maximum value of the mean velocity decreases when the overlap increases. As an illustration, it is remarked that, for the different values of the overlap equal to $(e\text{-}e')/d = 0$, $(e\text{-}e')/d = 0.1$, $(e\text{-}e')/d = 0.24$, and $(e\text{-}e')/d = 0.3$, the mean velocity values are equal, respectively, to $V = 14.9$ m s^{-1} (Fig. 4a), to $V = 14.5$ m s^{-1} (Fig. 4b), to $V = 14.2$ m s^{-1} (Fig. 4c), and to $V = 13.9$ m s^{-1} (Fig. 4d).

3.3 *Static Pressure*

Figure 5 presents the distribution of the static pressure in the computational domain with a zoom around the Savonius wind rotor for different overlap values. From this figure, there seems to be a compression zone that appears upstream of the Savonius wind rotor and which is accentuated at the concave surface of the upper bucket and the convex surface of the lower bucket of the Savonius wind rotor. At the concave surface of the lower bucket, there is a rapid decrease in the static pressure values. However, a depression occurs at the convex surface of the upper bucket. The most important depression zones appear at the two attack zones of the upper bucket of the Savonius rotor. This depression zone extends downstream of the rotor up to the limits of the computational domain. By comparing these results together, it can be observed that the maximum value of the static pressure is obtained for an overlap equal to $(e\text{-}e')/d = 0.24$. For the different values of the overlap equal to $(e\text{-}e')/d = 0$, $(e\text{-}e')/d = 0.1$, $(e\text{-}e')/d = 0.24$, and $(e\text{-}e')/\text{d} = 0.3$, the maximum values of the static pressure are equal, respectively, to $p = 20.5$ Pa (Fig. 5a), to $p = 33.4$ Pa (Fig. 5b), to $p = 145$ Pa (Fig. 5c), and to $p = 97.9$ Pa (Fig. 5d).

3.4 *Dynamic Pressure*

Figure 6 presents the distribution of the dynamic pressure in the computational domain with a zoom around the Savonius wind rotor for different overlap values. According to these results, it is worth noting that the dynamic pressure is quite low at the entry of the computational domain. At the two buckets, the value decreases and reaches zero. Furthermore, the occurrence of breakpoints has been remarked on the convex surface of the lower bucket and the concave surface of the upper bucket. However, at the external attack zones of the two buckets, there is a rapid increase in the dynamic pressure values. From these two zones, a wake characteristic of the maximum values of the dynamic pressure is developed and extends to the limits of the computational domain. Behind the rotor, there is a rapid decrease in the dynamic pressure. Comparing these results together, the internal overlap is

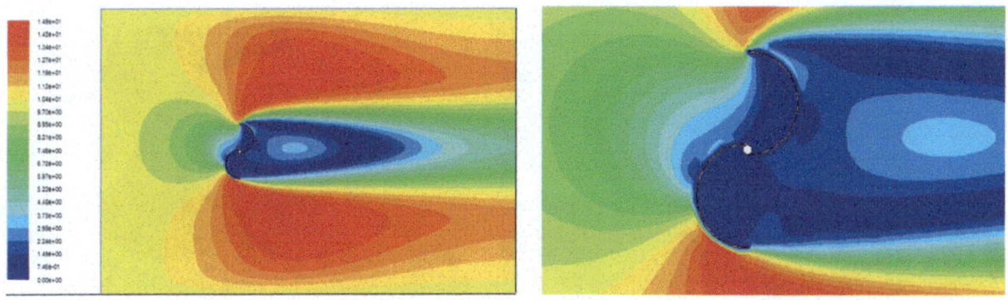

(a) (e-e')/d=0

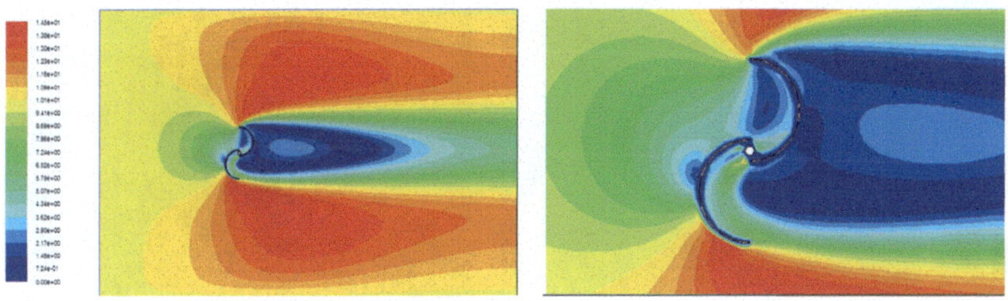

(b) (e-e')/d=0.1

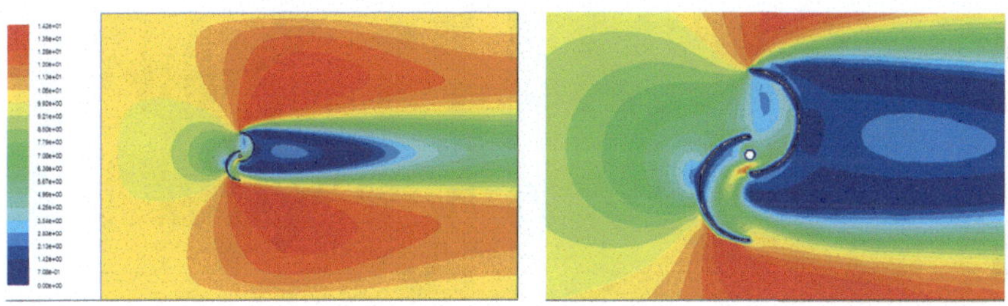

(c) (e-e')/d=0.24

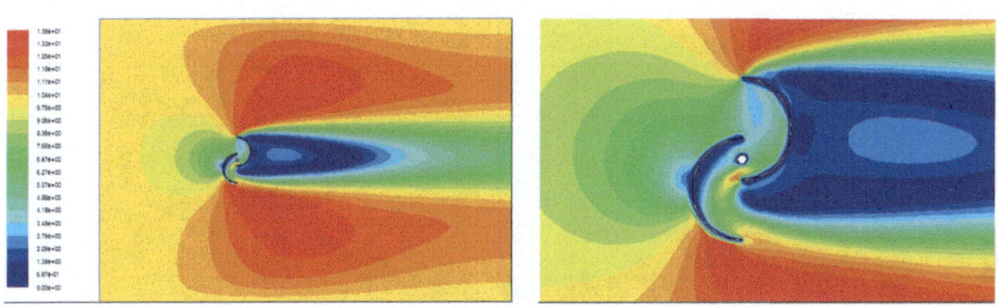

(d) (e-e')/d=0.3

Fig. 4 Distribution of the mean velocity

observed to have a direct effect on the dynamic pressure. In fact, in the internal attack zone and in the case of nonzero overlap ($(e\text{-}e')/d \neq 0$), there is a rapid increase in the dynamic pressure values. In addition, the maximum value of the dynamic pressure was noted to decrease when the overlap increases. Indeed, for the

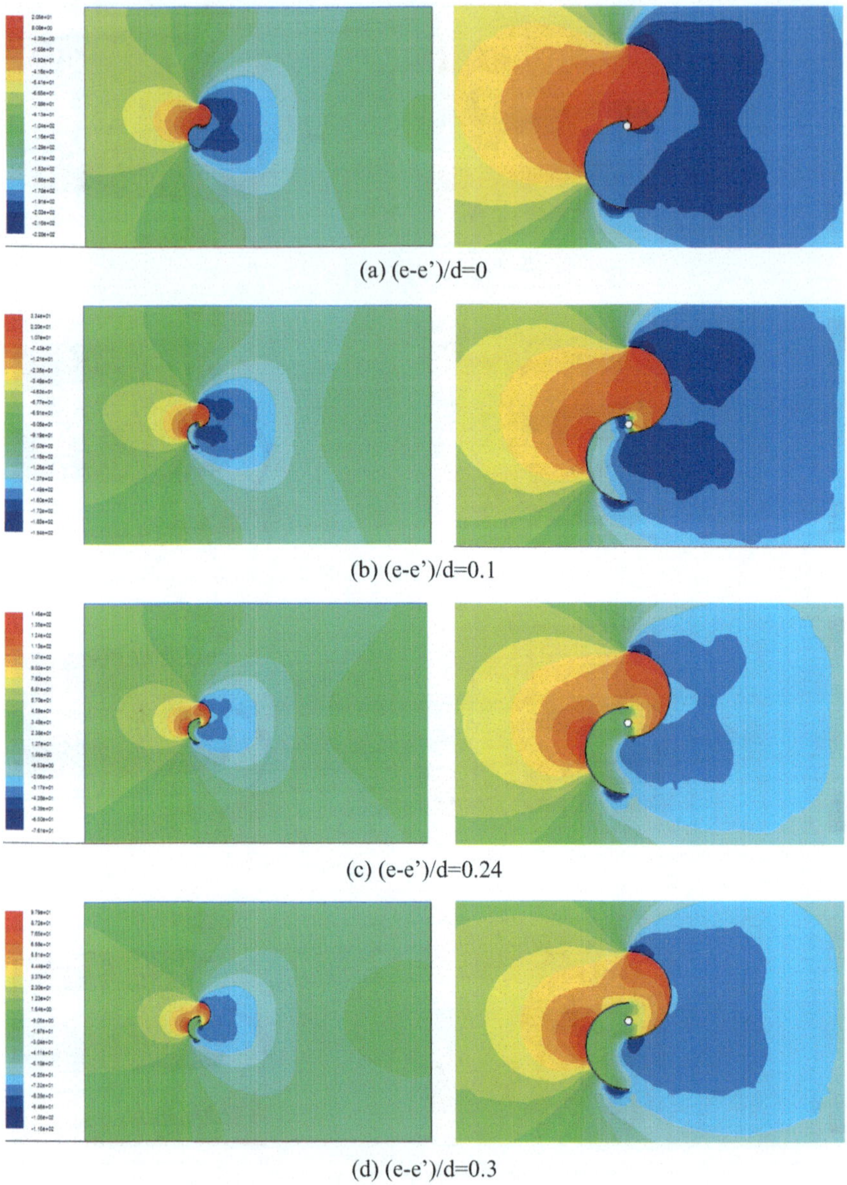

Fig. 5 Static power distribution

different values of the overlap equal to $(e\text{-}e')/d = 0$, $(e\text{-}e')/d = 0.1$, $(e\text{-}e')/d = 0.24$, and $(e\text{-}e')/d = 0.3$, the maximum values of the dynamic pressure are equal, respectively, to $p = 155$ Pa (Fig. 6a), to $p = 145$ Pa (Fig. 6b), to $p = 134$ Pa (Fig. 6c), and to $p = 130$ Pa (Fig. 6d).

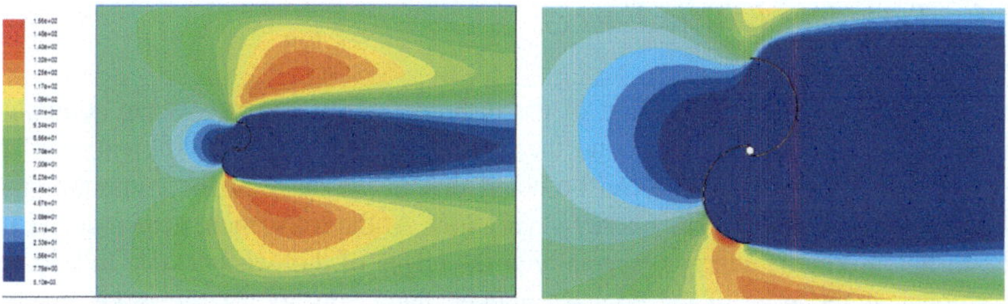

(a) (e-e')/d=0

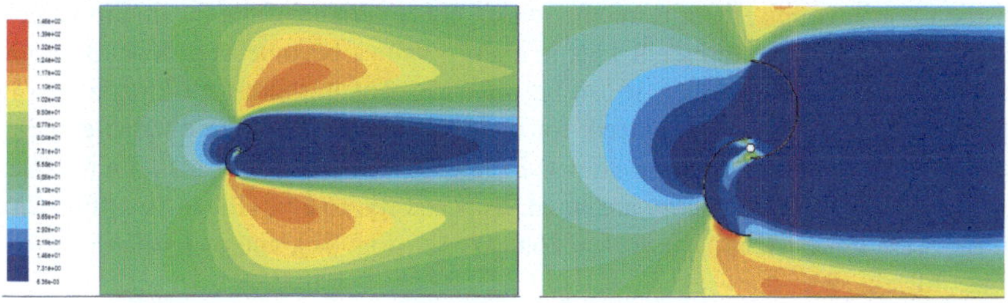

(b) (e-e')/d=0.1

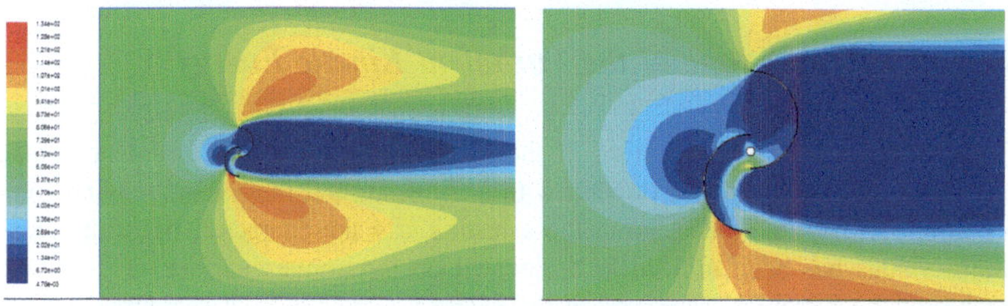

(c) (e-e')/d=0.24

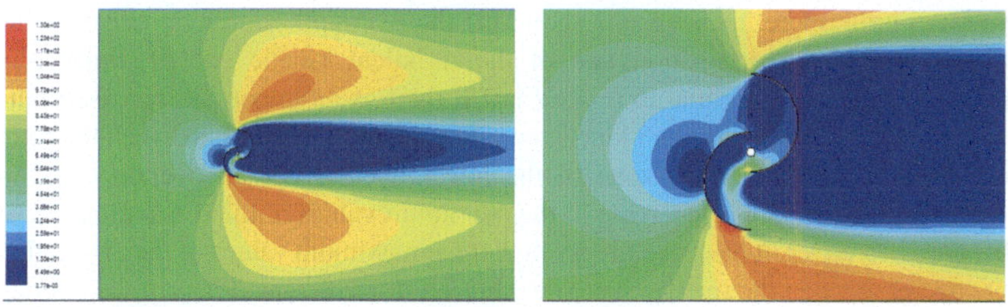

(d) (e-e')/d=0.3

Fig. 6 Dynamic pressure distribution

3.5 Turbulent Kinetic Energy

Figure 7 presents the distribution of the turbulent kinetic energy in the computational domain with a zoom around the Savonius wind rotor for different overlap values. From these results, it can be noted that in the entry of the computational domain, the turbulent kinetic energy is very low. At the rotor, there is an increase in the turbulent kinetic energy, especially at the edges of the two buckets. Indeed, in the upper attack zone of the concave bucket, there appears a large wake characteristic of the maximum values of the turbulent kinetic energy. At the lower bucket of the rotor Savonius, there is another wake, characteristic of the maximum values. Behind the rotor, the turbulent kinetic energy values become very low. By comparing these results together, it can be observed that the turbulent kinetic energy decreases with the overlap. Indeed, for the different values of the overlap equal to $(e\text{-}e')/d = 0$, $(e\text{-}e')/d = 0.1$, $(e\text{-}e')/d = 0.24$, and $(e\text{-}e')/d = 0.3$, the maximum values of the turbulent kinetic energy are equal, respectively, to $k = 41.9\ \mathrm{m^2\ s^{-2}}$ (Fig. 7a), to $k = 31.9\ \mathrm{m^2\ s^{-2}}$ (Fig. 7b), to $k = 26.6\ \mathrm{m^2\ s^{-2}}$ (Fig. 7c), and to $k = 24.5\ \mathrm{m^2\ s^{-2}}$ (Fig. 7d).

3.6 Dissipation Rate of the Turbulent Kinetic Energy

Figure 8 presents the distribution of the dissipation rate of the turbulent kinetic energy in the computational domain with a zoom around the Savonius wind rotor for different overlap values. From these results, a wake characteristic of the maximum values has been noticed to be located on the convex surface of the lower bucket of the Savonius rotor, and in the external attack zone of the upper bucket. Outside, there is a very rapid decrease in the dissipation rate of the turbulent kinetic energy. Furthermore, these results show that the overlap has a direct effect on the distribution of the dissipation rate of the turbulent kinetic energy. Indeed, in the case of an overlap equal to $(e\text{-}e')/d = 0.1$, there is an important increase in the dissipation rate of the turbulent kinetic energy value in the air gap between the rotor axis and the interior attack zone of the lower bucket. For an overlap equal to $(e\text{-}e')/d = 0.24$, there is a decrease in the maximum values of the dissipation rate at this gap. Moreover, it has been observed that the maximum values of the dissipation rate decrease when the overlap increases. Indeed, for different values of the overlap equal to $(e\text{-}e')/d = 0$, $(e\text{-}e')/d = 0.1$, $(e\text{-}e')/d = 0.242$, and $(e\text{-}e')/d = 0.3$, the maximum values of the dissipation rate of the turbulent kinetic energy are equal, respectively, to $\varepsilon = 435{,}000\ \mathrm{m^2\ s^{-3}}$ (Fig. 8a), to $\varepsilon = 366{,}000\ \mathrm{m^2\ s^{-3}}$ (Fig. 8b), to $\varepsilon = 347{,}000\ \mathrm{m^2\ s^{-3}}$ (Fig. 8c), and to $\varepsilon = 324{,}000\ \mathrm{m^2\ s^{-3}}$ (Fig. 8d).

(a) (e-e')/d=0

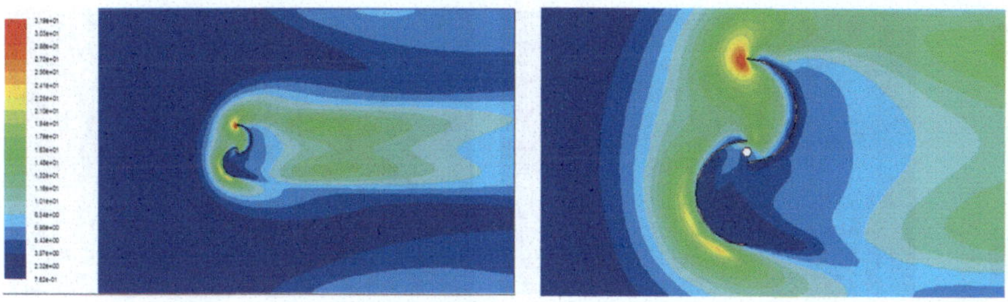

(b) (e-e')/d=0.1

(c) (e-e')/d=0.24

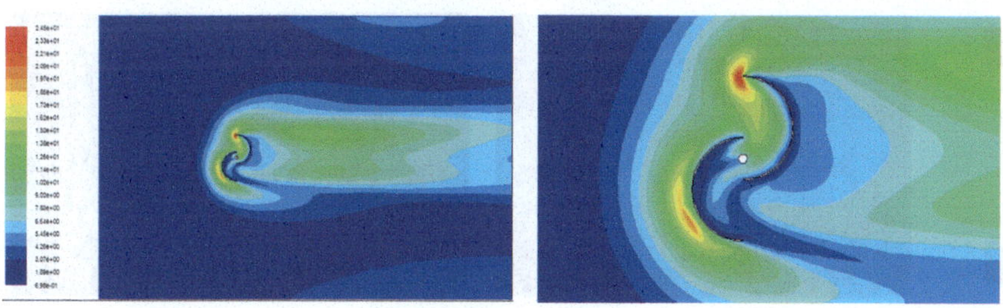

(d) (e-e')/d=0.3

Fig. 7 Turbulent kinetic energy distribution

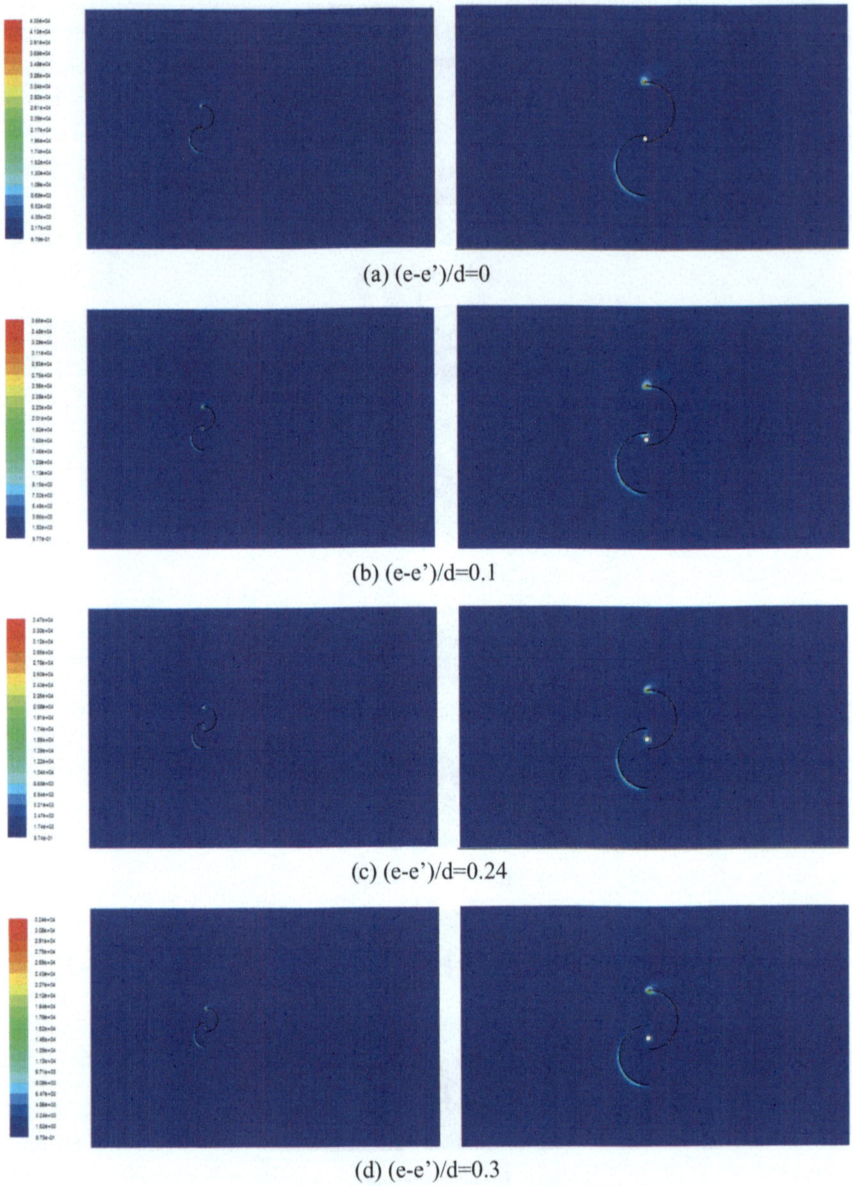

Fig. 8 Dissipation rate of the turbulent kinetic energy distribution

3.7 Turbulent Viscosity

Figure 9 presents the distribution of the turbulent viscosity in the computational domain with a zoom around the Savonius wind rotor for different overlap values. Based on these results, there is a resemblance between the different results. The turbulent viscosity has been noticed to be quite low upstream of the rotor. The values of the turbulent viscosity increase in the two upper and lower walls of the considered computational domain when two wake characteristics of the maximum values are developed. However, at the Savonius wind rotor, there is a rapid decrease in the turbulent viscosity values. In both upper and lower zones of the rotor, a wake characteristic of the low values of the turbulent viscosity is developed. This zone extends downstream of the rotor. In this case, there is a slight increase in the outlet section of the computational domain. Moreover, these results show that the distribution of the turbulent viscosity is affected by the overlap. In fact, the dissymmetry between the two wakes characteristic of the minimum values increases when the overlap increases. Upstream of the Savonius wind rotor, the distribution of the turbulent viscosity also depends on the value of the overlap. The highest values of the turbulent viscosity decrease when the overlap increases. Indeed, for the different values of the overlap equal to $(e\text{-}e')/d = 0$, $(e\text{-}e')/d = 0.1$, $(e\text{-}e')/d = 0.242$, and $(e\text{-}e')/d = 0.3$, the maximum values of the turbulent viscosity are equal, respectively, to $\mu_t = 0.289$ kg m^{-1} s^{-1} (Fig. 9a), to $\mu_t = 0.259$ kg m^{-1} s^{-1} (Fig. 9b), to $\mu_t = 0.229$ kg m^{-1} s^{-1} (Fig. 9c), and to $\mu_t = 0.219$ kg m^{-1} s^{-1} (Fig. 9d).

4 Comparison with Previous Results

In this section, we are interested in the study of the influence of the overlap on the variation of the static torque coefficient C_{Ms} of the Savonius wind rotor. In Fig. 10, the numerical results are compared with those obtained by Menet and Cottier (2003). According to these results, the maximum value of the static torque coefficient is obtained for an overlap equal to $e/d = 0.24$. This value is close to the value 0.242 found by Menet and Cottier (2003). Furthermore, the profile of the static torque presents the same evolution of the curve. The good agreement confirms the validity of the numerical method.

5 Conclusion

The numerical simulation of the turbulent flow around a Savonius wind rotor was investigated for different overlap values. According to the obtained results, the overlap has a direct effect on the turbulent flow. Local characteristics such as

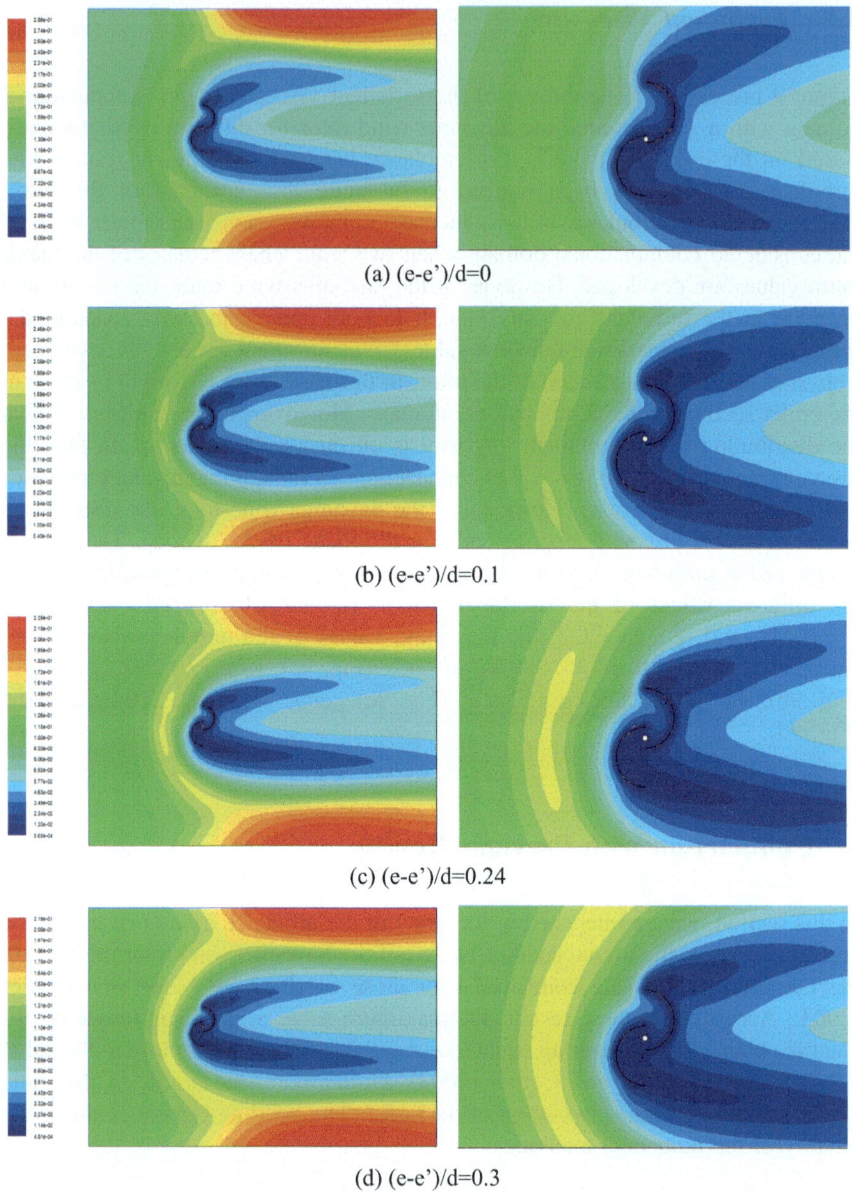

Fig. 9 Turbulent viscosity distribution

velocity field, mean velocity, static pressure, dynamic pressure, and turbulent characteristics differ from one configuration to another. The variation of the static torque coefficient of the Savonius wind rotor was also studied, and the numerical results were compared with those obtained in previous studies. A good agreement

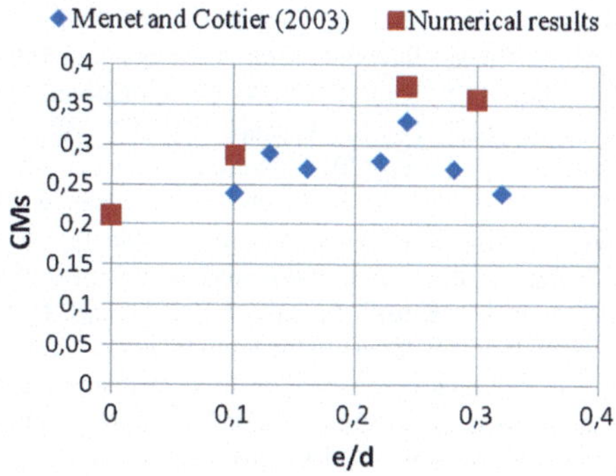

Fig. 10 Static torque coefficient CMs evolution

was obtained and confirmed the numerical method. In the future, we propose to study other geometrical parameters to improve the axial wind rotor operating conditions.

Nomenclature

ε	Turbulent kinetic energy dissipation rate, W kg^{-1}
μ	Dynamic viscosity, Pa s
μ_t	Turbulent viscosity, Pa s
ρ	Density, kg m^{-3}
σ_k	Constant of the k-ε turbulence model
σ_ε	Constant of the k-ε turbulence model
C_p	Coefficient of the power, dimensionless
$C_{1\varepsilon}$	Constant of the k-ε turbulence model, dimensionless
$C_{2\varepsilon}$	Constant of the k-ε turbulence model, dimensionless
C_μ	Constant of the k-ε turbulence model, dimensionless
F_i	Force components, N
G_k	Turbulence production term, kg m^{-1} s^{-3}
d	Rotor diameter, m
e	Bucket thickness, m
k	Turbulent kinetic energy, J kg^{-1}
P	Pressure, Pa
u_i	Velocity components, m.s^{-1}
u_i'	Fluctuating velocity components, m s^{-1}

References

Akwa JV, Júnior GA, Petry AP (2012) Discussion on the verification of the overlap ratio influence on performance coefficients of a Savonius wind rotor using computational fluid dynamics. Renew Energy 38:141–149

Aldos TK (1984) Savonius rotor using swinging blades as an augmentation system. Wind Eng 8:214–220

Blackwell BF, Sheldahl RE, Feltz LV (1978) Wind Tunnel performance data for two and three-bucket Savonius rotor. J Energy 2–3:160–164

Choudhury BJ, Saraf G (2014) Computational analysis of flow around a two-bladed Savonius rotor. ISESCO J Sci Technol 10(17):39–48

Driss Z, Abid MS (2012) Numerical investigation of the aerodynamic structure flow around Savonius wind rotor. Sci Acad Trans Renew Energy Syst Eng Technol 2(2):196–204

Driss Z, Mlayeh O, Driss S, Driss D, Maaloul M, Abid MS (2015) Study of the bucket design effect on the turbulent flow around unconventional Savonius wind rotors. Energy 89:708–729

Grinspan AS, Kumar S, Saha UK, Mahanta P, Ratnarao DV, Veda Bhanu G (2001) Design, development & testing of Savonius wind turbine rotor with twisted blades. In: Proceedings of international conference on fluid mechanics & fluid power, India, vol 28, pp 28–31

Kamoji MA, Kedare SB, Prabhu SV (2009) Experimental investigations on single stage modified Savonius rotor. Appl Energy 86:1064–1073

Khan N, Tariq IM, Hinchey M, Masek V (2009) Performance of Savonius rotor as water current turbine. J Ocean Technol 4(2):27–29

Menet JL, Bourabaa N (2004) Increase in the Savonius rotors efficiency via a parametric investigation. In: European wind energy conference, London

Menet JL, Cottier F (2003) Étude paramétrique du comportement aérodynamique d'une éolienne lente à axe vertical de type Savonius. 16è Congrès Français de Mécanique, Nice

Rogowski K, Maronski R (2015) CFD computation of the Savonius rotor. J Theor Appl Mech 53 (1):37–45

Saha UK, Rajkumar M (2005) On the performance analysis of Savonius rotor with twisted blades. J Renew Energy 960–1481

Sharma KK, Gupta R (2013) Flow field around three bladed Savonius rotor. Int J Appl Eng Res 8 (15):1773–1782. ISSN 0973-4562 © Research India Publications

Shikha, Bhatti TS, Kothari DP (2003) Wind energy conversion systems as a distributed source of generation. J Energy Eng 129(3):69–80

Ushiyama I, Nagai H (1988) Optimum design configurations and performances of Savonius rotors. Wind Eng 12–1:59–75

Study of the Collector Diameter Effect on the Characteristics of the Solar Chimney Power Plant

Ahmed Ayadi, Abdallah Bouabidi, Zied Driss and Mohamed Salah Abid

1 Introduction

Solar chimney power plant (SCPP) is a device designed for generating green energy. The SCPP uses sunlight and the natural buoyancy of heated air to harness energy from the sun. The air flow is driven by a pressure difference in the chimney; then, it will rotate the turbines inside the chimney to generate electricity. An SCPP makes significant contributions to the energy generation in countries where sunlight is available. The SCPP optimization requires the optimization of its main components such as the chimney, the collector, and the absorber. In the last few decades, a great deal of research was carried out on the SCPP optimization. The number of studies, focusing on the numerical methods to optimize the SCPP performance, is still on the increase. The first SCPP prototype was built in Manzanares, with 194.6 m high and 122 m radius. Haaf et al. (1983) published the results of preliminary tests such as energy audits, the collector efficiency values, pressure losses due to friction, and losses in the turbine section. Bernardes et al. (2003) examined the effect of various environmental conditions and structural dimensions of the solar chimney outputs. The results show that the chimney's height, the pressure drop factor across the turbine, the diameter, and the optical properties of the collector are

A. Ayadi (✉) · A. Bouabidi · Z. Driss · M. S. Abid
Laboratory of Electro-Mechanic Systems (LASEM), National Engineering School of Sfax (ENIS), University of Sfax (US), B.P. 1173 Km 3.5 Soukra, 3038 Sfax, Tunisia
e-mail: ahmed.ayadi.gem@gmail.com

A. Bouabidi
e-mail: bouabidi_abdallah@yahoo.fr

Z. Driss
e-mail: zied.Driss@enis.rnu.tn

M. S. Abid
e-mail: mohamedsalah.Abid@enis.rnu.tn

Z. Driss et al. (eds.), *CFD Techniques and Thermo-Mechanics Applications*,
https://doi.org/10.1007/978-3-319-70945-1_9

important parameters for the design of an SCPP. The production increases by increasing the height of the chimney and the collector diameter. Koonsrisuk et al. (2010) studied the generation of the form and the structure of the solar chimney in the pursuit of the overall performance of the flow system. The maximum mass flow rate and the maximum flow capacity were determined on the basis of a simple model. They found that the maximum flow capacity depends on the length scale of the plant. Bigger plants deliver more power per unit of domain. Zhou et al. (2009) analyzed the maximum chimney height for convection avoiding negative buoyancy at the chimney and the optimal chimney height for a maximum power output. They validated their work using a theoretical model with the measurements of the prototype in Manzanares. Maia et al. (2009) evaluated the influence of the geometrical parameters and material on the behavior of the air flow inside of the solar chimney. The analysis has proven that the geometric variables such as the diameter and the height of the tower directly influence the solar chimney's design. An increase in the diameter and in the tower height produces an increase in the mass flow rate. Ghalamchi et al. (2016) presented an experimental study on the thermal performance of a solar chimney with different parameters. The geometrical parameters and the absorbing material were analyzed. The results showed that aluminum is the best material as an absorber due to its better ability in heat transfer than iron plate because of its high thermal conductivity. Ghalamchi et al. (2013) constructed an SCPP with 3 m collector diameter and 2 m chimney height at the University of Tehran, Iran. In this work, they evaluated the temperature distribution and the air velocity in different geometrical cases. Alibakhsh et al. (2014) presented two-dimensional axisymmetric numerical simulations to analyze the effects of geometrical parameters on a constructed SCPP. Their results demonstrated that the chimney diameter has a greater influence than the chimney height. Hamdan (2013) showed that the chimney height, the collector radius, the turbine head, and the solar irradiance are essential parameters for the design of an SCPP. Gholamalizadeh and Kim (2014) analyzed the buoyancy-driven flow field and heat transfer inside the SCPP by simulating the greenhouse effect. Li et al. (2012) considered the heat and flow losses effects, the temperature lapse rate inside and outside the chimney. They have demonstrated that the foundation of the turbine in the SCPP system reduces the power output. According to these previous research works, it is clear that the study of the thermal systems using CFD methods has witnessed a great deal of development. In this chapter, we were interested in the study of the collector size effect on the performance of the SCPP system. The numerical model was validated by the experimental data of the Kasaeian et al. (2014) prototype.

2 Geometric Modeling

Figure 1 presents the geometric modeling studied in this chapter. The setup was built in the city of Tehran, Iran. Tehran has geographical length and width of 51.4° and 35.7°, respectively. The chimney height and diameter are equal to $H = 2$ m and

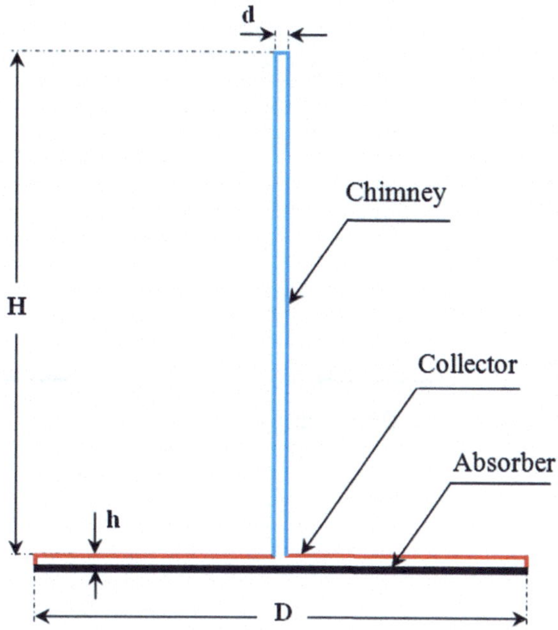

Fig. 1 Geometrical arrangements

$d = 0.2$ m, respectively, and the collector height and diameter are $h = 0.06$ m and $D = 3$ m, respectively. The considered system is similar to Kasaeian et al. (2014) application. Four collector diameters equal to $D = 100$ cm, $D = 200$ cm, $D = 300$ cm, and $D = 400$ cm have been studied in order to investigate the collector diameter effect on the SCPP performance. The different cases are presented in Fig. 2.

3 Numerical Method

A great deal of research has been numerically developed using the commercial CFD code ANSYS Fluent. The mathematical equations governing the air flow inside the SCPP are the continuity equation, the momentum equations, and the energy equation (Driss et al. 2014, 2015; Frikha et al. 2015).

A standard atmospheric pressure and the temperature conditions are imposed. The collector inlet is specified as the pressure inlet, and the chimney outlet is specified as the pressure outlet. Constant temperatures are applied to the walls, and the chimney wall is assumed insulated. The ambient temperature is 306 K, and the air is an incompressible ideal gas. The collector covering is considered as a semitransparent wall. The boundary conditions are given in Table 1.

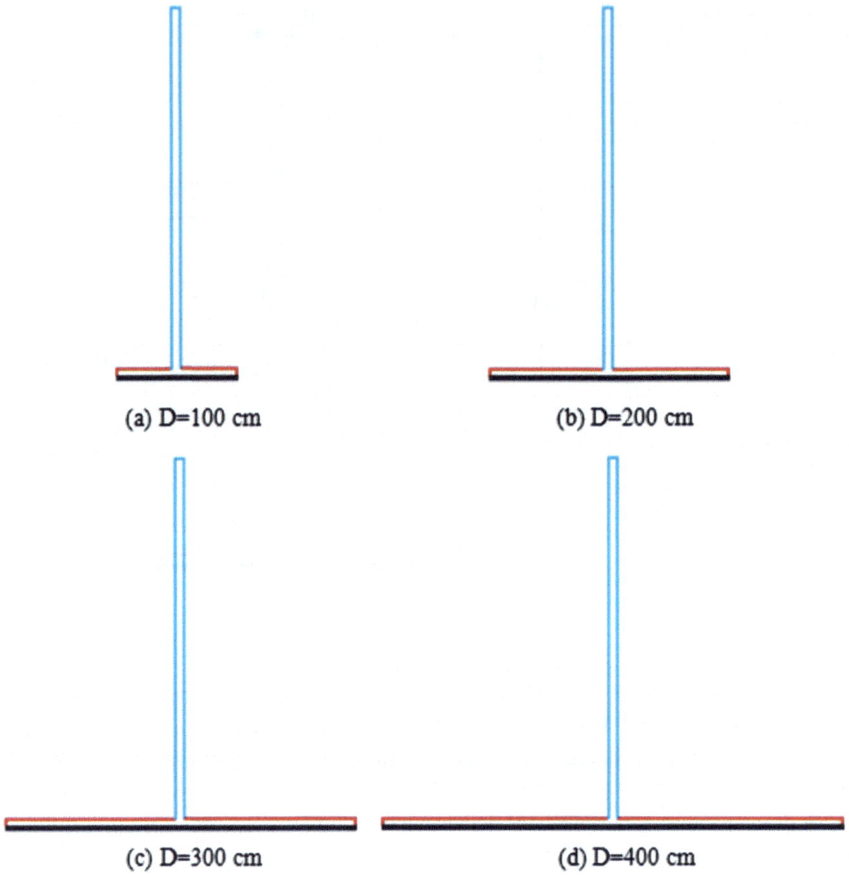

Fig. 2 Different configurations with different collector diameters

Table 1 Boundary conditions

Surface	Type	Value
Collector inlet	Pressure inlet	ΔP = 0 Pa and T = 306 K
Chimney outlet	Pressure outlet	ΔP = 0 Pa and T = 306 K
Chimney wall	Opaque wall	$q = 0$ W m^{-2}
Collector	Semitransparent wall	$q = 800$ W m^{-2}

4 Numerical Results

The obtained results here are expected to provide information that will help improve the overall efficiency of the SCPP.

4.1 Temperature

Figure 3 depicts the distribution of the temperature in the whole SCPP system volume for the different considered diameters—D = 100 cm, D = 200 cm, D = 300 cm, and D = 400 cm. According to these results, it can be noted that the temperature distribution is similar for the different considered diameter sizes. The maximum value location is the same for the different considered cases, and it is located near the absorber. This might be due to the heat transfer exchange between the absorber and the airflow inside the collector. In fact, it can be explained by the mass conservation. However, the minimum value is located in the inlet of the

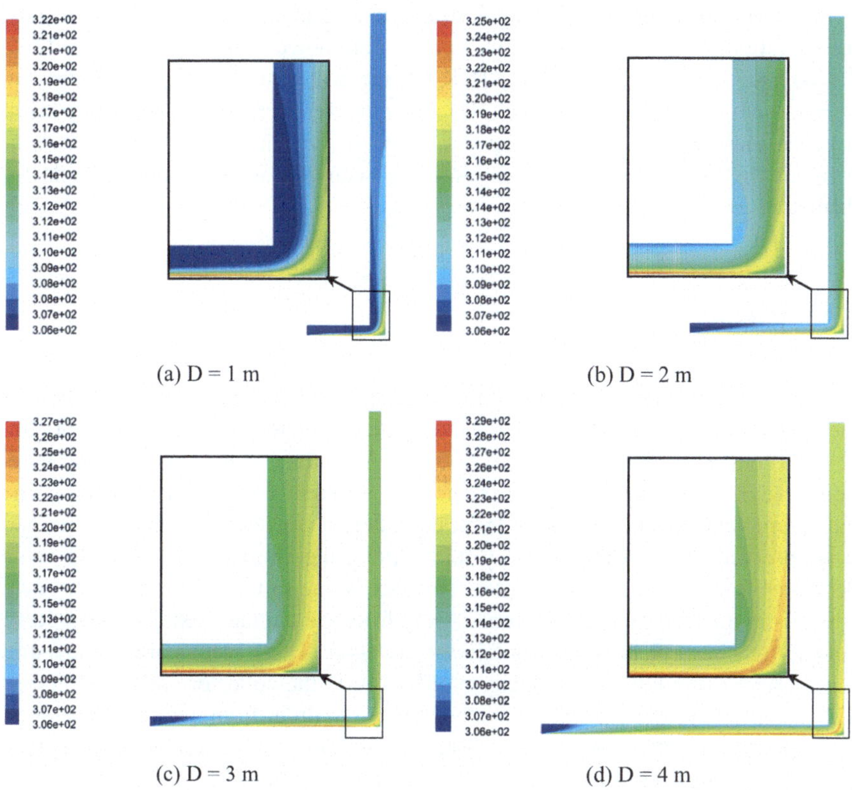

Fig. 3 Temperature distribution

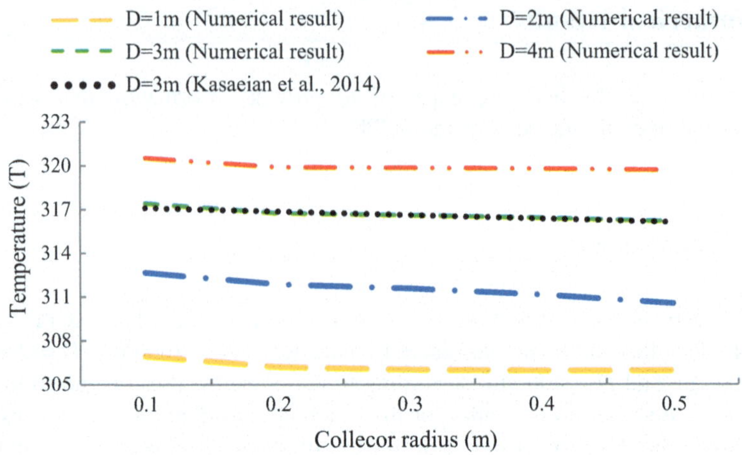

Fig. 4 Profiles of the temperature along the collector radius

collector. Figure 4 shows the temperature profiles inside the collector. A difference in the maximum value was also noted from one case to another. From these results, the maximum temperatures inside the collector are noted to be equal to $T = 307$ K, $T = 313$ K, $T = 317$ K, and $T = 321$ K for the collector diameters $D = 100$ cm, $D = 200$ cm, $D = 300$ cm, and $D = 400$ cm, respectively. The comparison between these results confirms that the collector size has a direct effect on temperature. In fact, the temperature increases with the increase in the collector diameter.

4.2 Velocity

Figure 5 presents the distribution of the magnitude velocity in the whole volume of the SCPP system for the different considered diameter sizes—$D = 100$ cm, $D = 200$ cm, $D = 300$ cm, and $D = 400$ cm. According to these results, the magnitude velocity is noted to be similar for the different considered diameters. The maximum value location is the same for the different considered configurations, and it is observed to be in the chimney inlet. Along the chimney, the velocity value decreases from the inlet to the outlet. This fact is due to the difference of pressure which is at its maximum in this zone. Away from the chimney axis, the magnitude velocity reaches a very weak value in the collector. These results are displayed in Figs. 6 and 7 providing the velocity profiles inside the collector and the chimney. The magnitude velocity maximum value, it noted to differ from one case to another. For $D = 100$ cm, $D = 200$ cm, $D = 300$ cm, and $D = 400$ cm collector diameters, the maximum velocities are, respectively, equal to $V = 1.76$ m s^{-1}, $V = 1.93$ m s^{-1}, $V = 2.12$ m s^{-1}, and $V = 2.22$ m s^{-1}. The comparison between

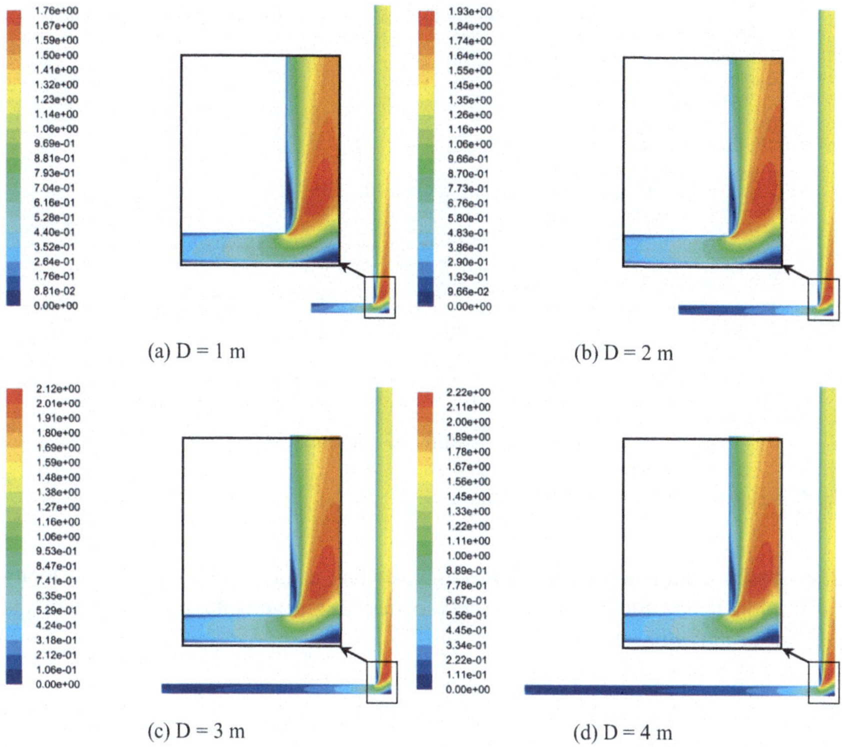

Fig. 5 Distribution of the magnitude velocity

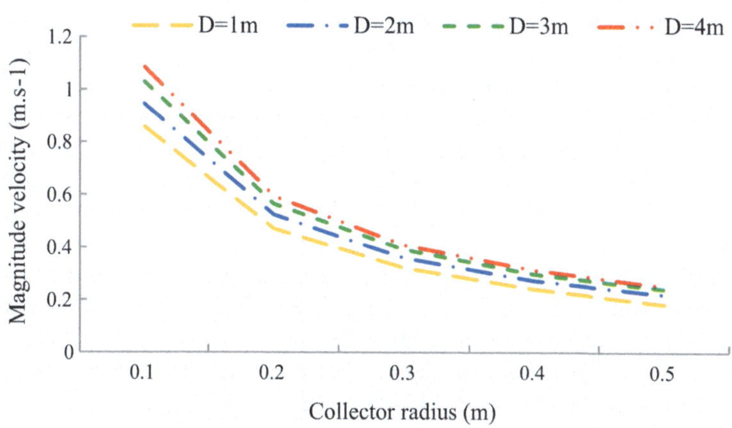

Fig. 6 Magnitude velocity profiles along the collector radius

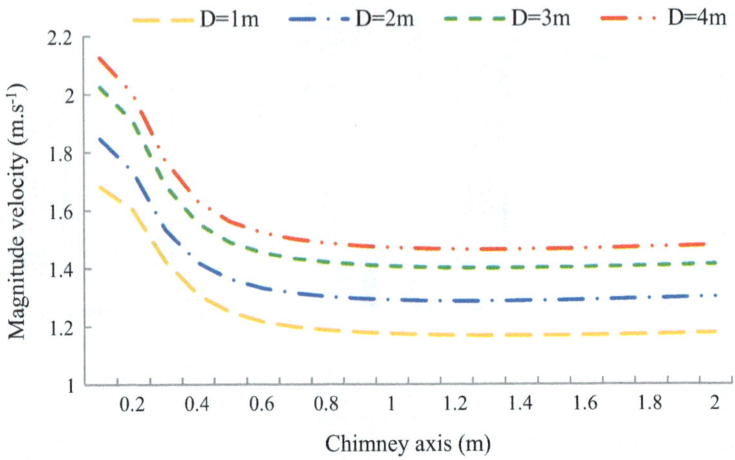

Fig. 7 Velocity profiles along the chimney axis

these results confirms that the collector size has a direct effect on the velocity distribution inside the SCPP. In fact, the magnitude velocity increases with the increase in the collector diameter.

4.3 *Dynamic Pressure*

Figure 8 depicts the dynamic pressure distribution in the whole volume of the SCPP system for the different considered diameters—D = 100 cm, D = 200 cm, D = 300 cm, and D = 400 cm. According to these results, the dynamic pressure distribution conserves the same distribution for the different considered diameter sizes. Along the collector radius, the dynamic pressure is globally uniform and reaches a very weak value. Indeed, a compression zone is recorded at the chimney inlet close to the axis. This fact is explained by the air velocity which reaches its maximum in this zone. The static pressure maximum value varies from one configuration to another. The comparison between these results confirms that the diameter dimension has a direct effect on the dynamic pressure inside the SCPP. In fact, the values of the dynamic pressure increase with the increase in the collector diameter. For the collector diameters D = 100 cm, D = 200 cm, D = 300 cm, and D = 400 cm, the maximum dynamic pressures are, respectively, equal to p_d = 1.74 Pa, p_d = 2.09 Pa, p_d = 2.52 Pa, and p_d = 2.77 Pa.

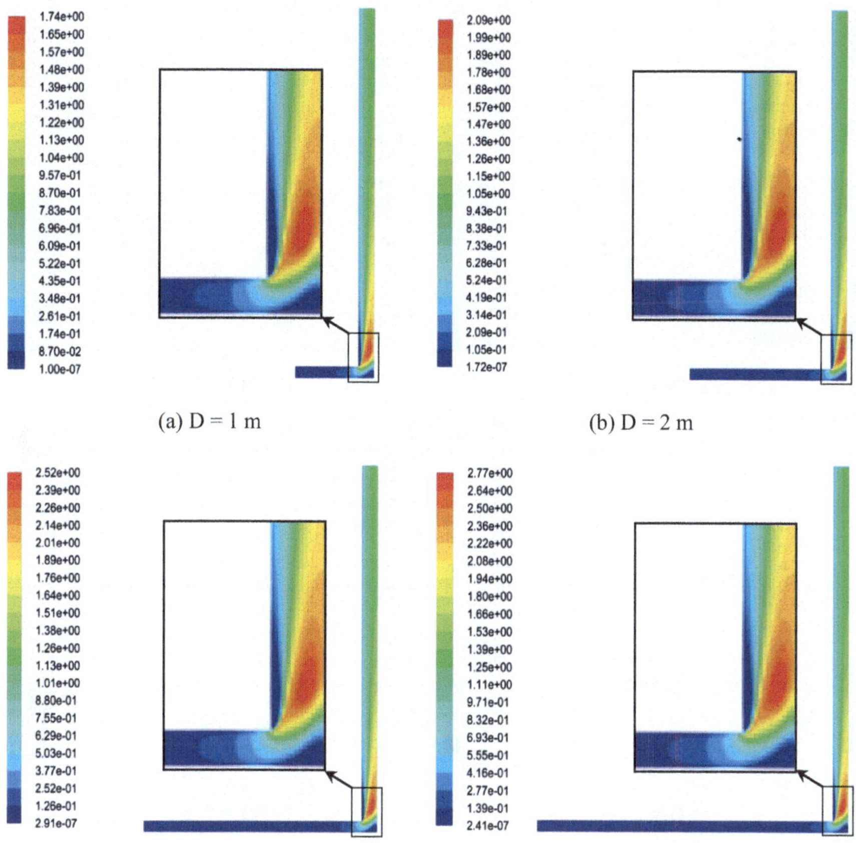

(a) D = 1 m (b) D = 2 m

(c) D = 3 m (d) D = 4 m

Fig. 8 Dynamic pressure distribution

4.4 Total Pressure

Figure 9 depicts the distribution of the total pressure in the whole volume of the SCPP system for the different considered diameters (D = 100 cm, D = 200 cm, D = 300 cm, and D = 400 cm). From these results, the total pressure distribution is the same for the different considered cases. The compression zone appears at the chimney outlet defined by z = 3 m, and it extends over the chimney axis until z = 2.5 m. However, the depression zone appears at the chimney inlet near the wall for all the considered diameters. Beyond these areas, the values of the total pressure vary slightly. The maximum value of the total pressure varies from one configuration to another. The comparison between these results confirms that the diameter dimension has a direct effect on the total pressure inside the SCPP. In fact, the total pressure values increase with the increase in collector diameter. For D = 100 cm,

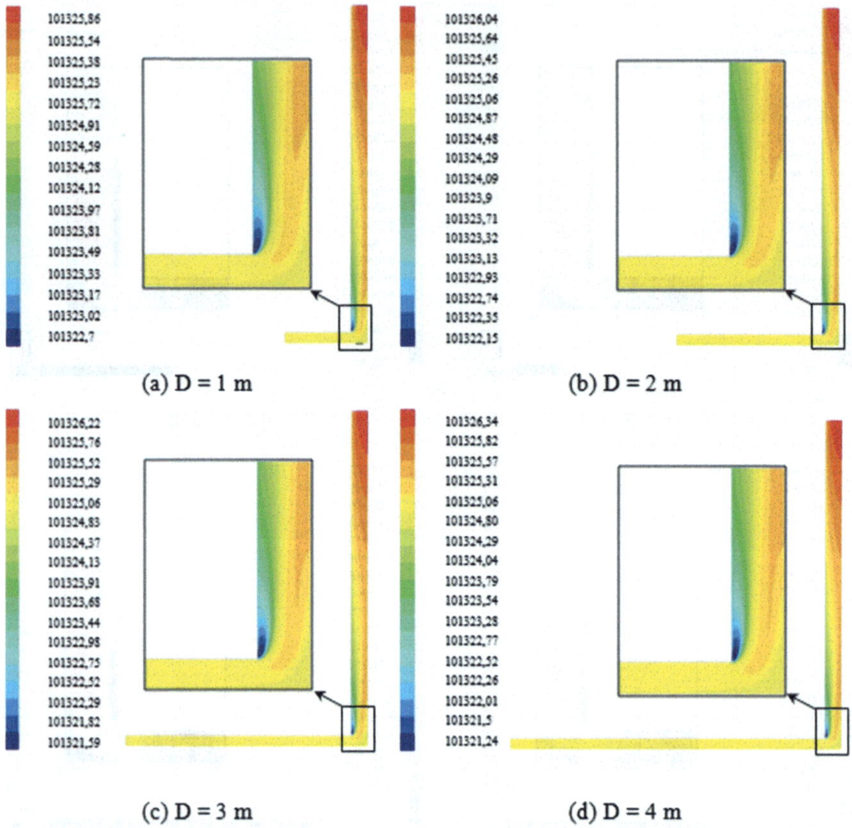

Fig. 9 Total pressure distribution

D = 200 cm, D = 300 cm, and D = 400 cm collector diameters, the maximum velocities are, respectively, equal to p = 101325.86 Pa, p = 101326.04 Pa, p = 101326.22 Pa, and p = 101326.34 Pa.

4.5 Turbulent Kinetic Energy

Figure 10 presents the distribution of the turbulent kinetic energy in the whole volume of the SCPP system for the different considered diameters—D = 100 cm, D = 200 cm, D = 300 cm, and D = 400 cm. According to these results, it has been noted that the distribution of the turbulent kinetic energy is similar for the different considered diameter sizes. Indeed, the wake characteristic of the maximum values appears in the same location for the different considered cases. In fact, it has been observed in the inlet of the chimney near to the wall. The maximum value is noted

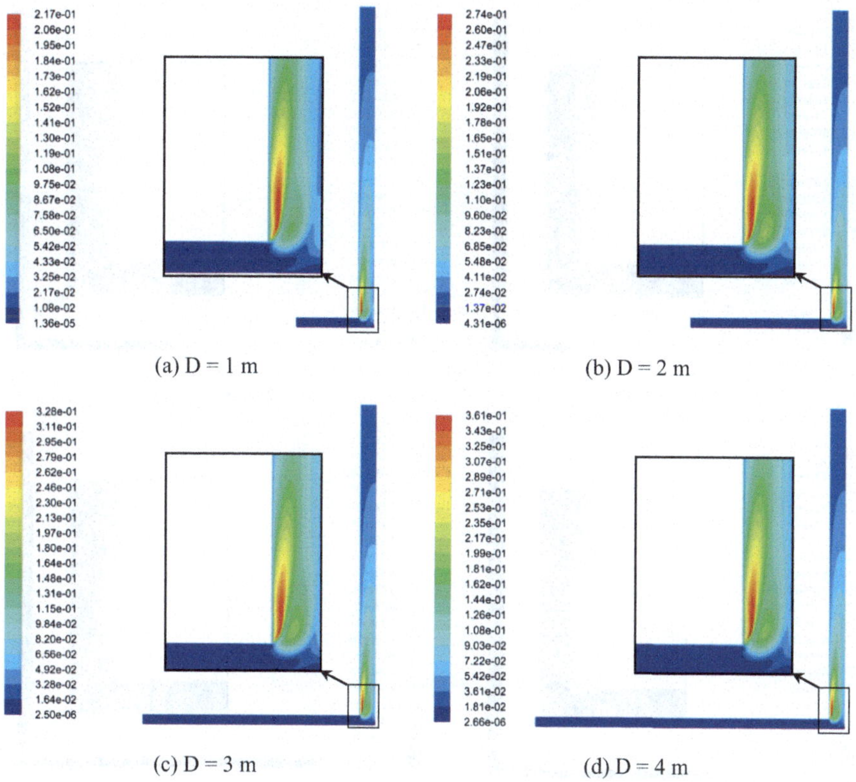

Fig. 10 Turbulent kinetic energy distribution

to differ from one case to another. The comparison between these results confirms that the diameter size has a direct effect on the maximum value of the turbulent kinetic energy. In these conditions, the maximum value of the turbulent kinetic energy is equal to $k = 0.217\ \mathrm{m}^2\ \mathrm{s}^{-2}$, $k = 0.274\ \mathrm{m}^2\ \mathrm{s}^{-2}$, $k = 0.328\ \mathrm{m}^2\ \mathrm{s}^{-2}$, and $k = 0.361\ \mathrm{m}^2\ \mathrm{s}^{-2}$ for the diameters equal to $D = 100$ cm, $D = 200$ cm, $D = 300$ cm, and $D = 400$ cm, respectively.

4.6 *Turbulent Kinetic Energy Dissipation Rate*

Figure 11 shows the dissipation rate distribution of the turbulent kinetic energy in the whole volume of the SCPP system for the different considered diameters ($D = 100$ cm, $D = 200$ cm, $D = 300$ cm, and $D = 400$ cm). According to these results, the dissipation rate distribution of the turbulent kinetic energy is noted to be similar for the different considered diameter sizes. It is clear that the turbulent kinetic energy dissipation rate is very weak all over the system expect at the

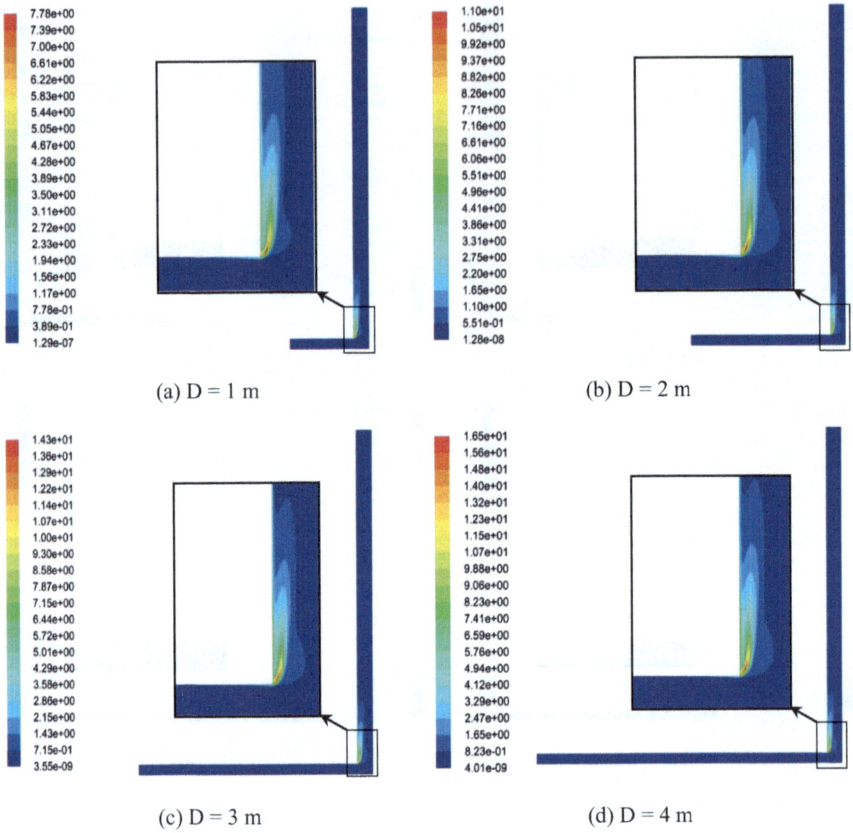

Fig. 11 Turbulent dissipation rate distribution

chimney inlet near the wall. Indeed, the maximum value location is the same for the different considered configurations. It has been noted that the maximum value varies from one configuration to another. The comparison between these results confirms that the diameter dimension has a direct effect on the turbulent dissipation rate distribution. For D = 100 cm, D = 200 cm, D = 300 cm, and D = 400 cm, the maximum value of the turbulent kinetic energy dissipation rate is $\varepsilon = 7.8$ W kg^{-1}, $\varepsilon = 11$ W kg^{-1}, $\varepsilon = 14.3$ W kg^{-1}, and $\varepsilon = 16.5$ W kg^{-1}, respectively.

4.7 Turbulent Viscosity

Figure 12 presents the distribution of the turbulent viscosity in the whole volume of the SCPP system for the different considered diameters (D = 100 cm, D = 200 cm, D = 300 cm, and D = 400 cm). According to these results, the turbulent viscosity

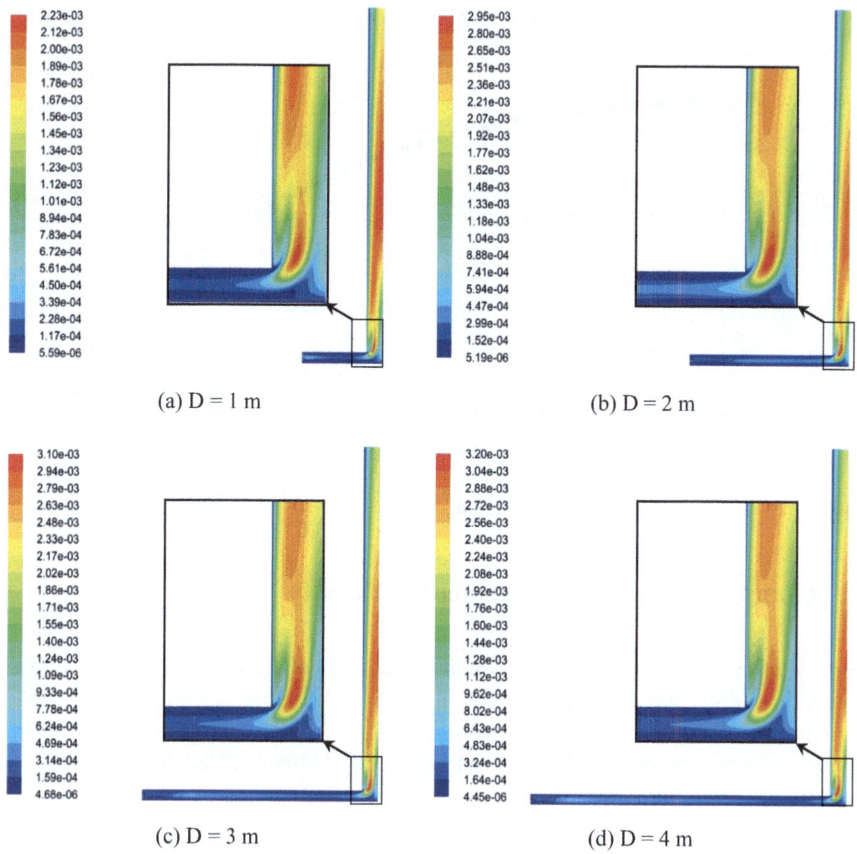

Fig. 12 Turbulent viscosity distribution

presents the same distribution for all the considered diameters. The turbulent viscosity is very weak in the collector but reaches its maximum at the chimney. This can be explained by the collision between the particles constituting the air volume. However, there is a slight difference between the maximum values of the turbulent kinetic energy for each diameter. For D = 100 cm, D = 200 cm, D = 300 cm, and D = 400 cm, the maximum value of the turbulent viscosity is, respectively, equal to $\mu_t = 0.00223$ kg m^{-1} s^{-1}, $\mu_t = 0.00295$ kg m^{-1} s^{-1}, $\mu_t = 0.00310$ kg m^{-1} s^{-1}, and $\mu_t = 0.0032$ kg.m^{-1} s^{-1}. The comparison between these results confirms that the diameter dimension has a direct effect on the turbulent viscosity. In fact, the turbulent viscosity increases with the increase in the collector diameter.

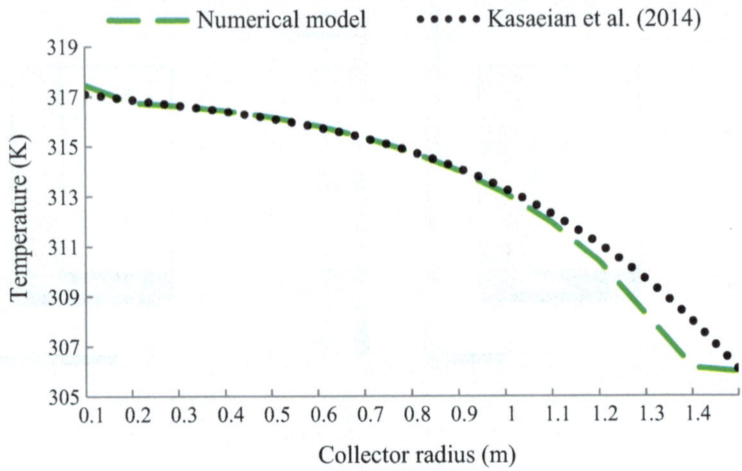

Fig. 13 Comparison with experimental results

5 Comparison with Experimental Results

Figure 13 shows a comparison between the numerical simulations and the experimental results reached by Kasaeian et al. (2014). According to this figure, a good agreement can be observed between numerical and experimental results measured by temperature sensors. The considered measurement line is situated at the collector center.

6 Conclusion

The objective of this chapter was to analyze the effect of the diameter size on the SCPP performance. To this end, a numerical simulation of the fluid flow inside an SCPP system was investigated using the commercial software "ANSYS Fluent 17.0." The numerical results confirm that the diameter dimensions have a direct effect on the local characteristics. As a future perspective, we are going to design and optimize an SCPP system for North Africa. We will investigate the effects of various geometrical parameters such as the chimney diameter and height.

Acknowledgements This work is partially supported by the Laboratory of Electro-Mechanic Systems (LASEM). The authors also gratefully acknowledge the reviewers' helpful comments and suggestions, which improved the chapter quality.

References

Bernardes MA dos S, Voss A, Weinrebe G (2003) Thermal and technical analyzes of solar chimneys. Sol Energy 75:511–524

Driss Z, Mlayeh O, Driss D, Maaloul M, Abid MS (2014) Numerical simulation and experimental validation of the turbulent flow around a small incurved Savonius wind rotor. Energy 74:506–551

Driss Z, Mlayeh O, Driss S, Driss D, Maaloul M, Abid MS (2015) Study of the bucket design effect on the turbulent flow around unconventional Savonius wind rotors. Energy 89:708–729. https://doi.org/10.1016/j.energy.2015.06.023

Frikha S, Driss Z, Hagui MA (2015) Computational study of the diffuser angle effect in the design of awaste heat recovery system for oil field cabins. Energy 84:219–238

Gholamalizadeh E, Kim M-H (2014) Three-dimensional CFD analysis for simulating the greenhouse effect in solar chimney power plants using a two-band radiation model. Renew Energy 63:498–506

Ghalamchi M, Kasaeian A, Ghalamchi M (2013) Experimental study of geometrical and climate effects on the performance of a small solar chimney. Renew Energy 56:50–54

Ghalamchi M, Kasaeian A, Ghalamchi M, Mirzahosseini AH (2016) An experimental study on the thermal performance of a solar chimney with different dimensional parameters. Renew Energy 91:477–483

Haaf W, Friedrich K, Mayr G, Schlaich J (1983) Solar chimneys. Part1: principle and construction of the pilot plant in Manzanares. Int J Sol Energy 2(1):3–20

Hamdan MO (2013) Analysis of solar chimney power plant utilizing chimney discrete model. Renew Energy 56:50–54

Kasaeian A, Ghalamchi M, Ghalamchi M (2014) Simulation and optimization of geometric parameters of a solar chimney in Tehran. Energy Convers Manage 83:28–34

Koonsrisuk A, Lorente S, Bejan A (2010) Constructal solar chimney configuration. Int J Heat Mass Transf 53:327–333

Li J-Y, Guo P-H, Wang Y (2012) Effects of collector radius and chimney height on power output of a solar chimney power plant with turbines. Renew Energy 47:21–28

Maia CB, Ferreira AG, Valle RM, Cortez MFB (2009) Theoretical evaluation of the influence of geometric parameters and materials on the behavior of the air flow in a solar chimney. Comput Fluids 38:625–636

Zhou X, Yang J, Xiao B, Hou G, Xing F (2009) Analysis of chimney height for solar chimney power plant. Appl Therm Eng 29:178–185

The manufacturer's authorised representative in the EU is Springer
Nature Customer Service Centre GmbH, Europaplatz 3, 69115 Heidelberg,
Germany. If you have any concerns regarding our products, please
contact ProductSafety@springernature.com

Printed and bound by CPI Group (UK) Ltd, Croydon, CR0 4YY
25/09/2026
02222836-0005